T0325902

Atlantis Studies in Dynamical Systems

Volume 4

The “Atlantis Studies in Dynamical Systems” publishes monographs in the area of dynamical systems, written by leading experts in the field and useful for both students and researchers. Books with a theoretical nature will be published alongside books emphasizing applications.

More information about this series at http://www.atlantis-press.com

Zhiqiang Li

Ergodic Theory of Expanding Thurston Maps

Zhiqiang Li
Institute for Mathematical Sciences
Stony Brook University
Stony Brook, NY
USA

Atlantis Studies in Dynamical Systems
ISBN 978-94-6239-173-4 ISBN 978-94-6239-174-1 (eBook)
DOI 10.2991/978-94-6239-174-1

Library of Congress Control Number: 2017930418

Printed on acid-free paper

To the loving memory of my grandmother
Fengxian Shen

Preface

This monograph came out of my thesis work under the supervision of my Ph.D. advisor Mario Bonk during my graduate studies at the University of Michigan, Ann Arbor, and later at the University of California, Los Angeles. It focuses on the dynamics, more specifically ergodic theory, of some continuous branched covering maps on the 2-sphere, called expanding Thurston maps.

More than 15 years ago, Mario Bonk and Daniel Meyer became independently interested in some basic problems on quasisymmetric parametrization of 2-spheres, related to the dynamics of rational maps. They joined forces during their time together at the University of Michigan and started their investigation of a class of continuous (but not necessarily holomorphic) maps modeling a subclass of rational maps. These maps belong to a bigger class of continuous maps on the 2-sphere studied by William P. Thurston in his famous characterization theorem of rational maps (see [DH93]). As a result, Mario Bonk and Daniel Meyer called their maps *expanding Thurston maps*. Related studies were carried out by other researchers around the same time, notably Peter Haïssinsky and Kevin Pilgrim [HP09], and James W. Cannon, William J. Floyd, and Walter R. Parry [CFP07].

By late 2010, Mario Bonk and Daniel Meyer had summarized their findings in a reader-friendly arXiv draft [BM10] entitled *Expanding Thurston maps*, which they initially intended to publish in the AMS Mathematical Surveys and Monographs series. In order to make the material even more accessible, they decided later to expand their draft. This led to a long delay for the final published version [BM17] with almost twice the size of [BM10].

I was introduced to expanding Thurston maps by Mario Bonk soon after I joined in the graduate program at the University of Michigan. I quickly got deeply fascinated by this subject due to the connections to geometry, analysis, combinatorics, and dynamical systems.

I finished my first project on the periodic points and properties of the measures of maximal entropy of expanding Thurston maps under the supervision of Mario Bonk (later appeared in [Li13], see Chap. 4) after we moved to Los Angeles. I then decided to continue working on the ergodic theory of expanding Thurston maps, on which I eventually wrote my thesis.

This monograph covers investigations on the measures of maximal entropy, and more generally, equilibrium states of expanding Thurston maps, and their relations to the periodic points and the preimage points. In order to study the equilibrium states, the theory of thermodynamical formalism for Hölder continuous potentials is established in our context (see Chap. 5). The study of equidistribution results also leads to a close investigation on the expansion properties of our dynamical systems (see Chap. 6) and the discovery of some large deviation results (see Chap. 7).

This monograph is also intended to serve as a basic reference for the theory of thermodynamical formalism in our context. The applications to the study of the dynamical zeta functions were also kept in mind when this monograph was being prepared. As such, complex-valued function spaces are used whenever they do not introduce too much complication.

Acknowledgments I want to express my deep gratitude to Mario Bonk for introducing me to this subject, and for his encouragements, guidance, and strict standards. I would also like to thank Ilia Binder, Nhan-Phu Chung, Peter Haïssinksy, Daniel Meyer, and Kevin Pilgrim for helpful conversations. My gratitude also goes to our editors Boris Hasselblatt, Henk W. Broer, and Arjen Sevenster for their patient support and helpful comments. I want to acknowledge the partial supports from NSF grants DMS-1162471 and DMS-1344959. Last but not least, I want to thank my wife, Xuan Zhang, for her understanding, support, and love.

Stony Brook, NY, USA and Hangzhou, China Zhiqiang Li
December 2016

Contents

Notation

Let $\mathbb{C}$ be the complex plane and $\widehat{\mathbb{C}}$ be the Riemann sphere. We use the convention that $\mathbb{N} = \{1, 2, 3, \ldots\}$, $\mathbb{N}_0 = \{0\} \cup \mathbb{N}$, and $\widehat{\mathbb{N}} = \mathbb{N} \cup \{+\infty\}$, with the order relations $<$, $\leq$, $>$, $\geq$ defined in the obvious way. As usual, the symbol log denotes the logarithm to the base e, and $\log_b$ the logarithm to the base b for $b > 0$.

The cardinality of a set A is denoted by card$A$. For $x \in \mathbb{R}$, we define $\lfloor X \rfloor$ as the greatest integer $\leq x$, and $\lfloor X \rfloor$ the smallest integer $\geq x$.

Let $g \colon X \to Y$ be a function between two sets X and Y. We denote the restriction of g to a subset Z of X by $g|_Z$.

Let (X, d) be a metric space. For subsets $A, B \subseteq X$, we set $d(A, B) = \inf\{d(x, y) \,|\, x \in A, y \in B\}$, and $d(A, x) = d(x, A) = d(A, \{x\})$ for $x \in X$. For each subset $Y \subseteq X$, we denote the diameter of Y by $\mathrm{diam}_d(Y) = \sup\{d(x, y) \,|\, x, y \in Y\}$, the interior of Y by $\mathrm{int}\, Y$, and the characteristic function of Y by $\mathbb{1}_Y$, which maps each $x \in Y$ to $1 \in \mathbb{R}$. We use the convention that $\mathbb{1} = \mathbb{1}_X$ when the space X is clear from the context. The identity map $\mathrm{id}_X \colon X \to X$ sends each $x \in X$ to x itself. For each $r > 0$, we define $N_d^r(A)$ to be the open r-neighborhood $\{y \in X \,|\, d(y, A) < r\}$ of A, and $\overline{N_d^r}(A)$ the closed r-neighborhood $\{y \in X \,|\, d(y, A) \leq r\}$ of A. For $x \in X$, we denote the open (resp. closed) ball of radius r centered at x by $B_d(x, r)$ (resp. $\overline{B_d}(x, r)$).

We set $C(X)$ (resp. $B(X)$) to be the space of continuous (resp. bounded Borel) functions from X to $\mathbb{R}$, by $\mathcal{M}(X)$ the set of finite signed Borel measures, and $\mathcal{P}(X)$ the set of Borel probability measures on X. We denote by $C(X, \mathbb{C})$ (resp. $B(X, \mathbb{C})$) the space of continuous (resp. bounded Borel) functions from X to $\mathbb{C}$. Obviously $C(X) \subseteq C(X, \mathbb{C})$ and $B(X) \subseteq B(X, \mathbb{C})$. We will adopt the convention that unless specifically referring to $\mathbb{C}$, we only consider real-valued functions.

For $\mu \in \mathcal{M}(X)$, we use $\|\mu\|$ to denote the total variation norm of μ, $\mathrm{supp}\mu$ the support of μ, and

$$\langle \mu, u \rangle = \int u \, \mathrm{d}\mu$$

for each $u \in C(S^2)$. If we do not specify otherwise, we equip $C(X)$ and $C(X,\mathbb{C})$ with the uniform norm $\|\cdot\|_\infty$. For a point $x \in X$, we define δ_x as the Dirac measure supported on $\{x\}$. For $g \in C(X)$ we set $\mathcal{M}(X,g)$ to be the set of g-invariant Borel probability measures on X. Unless otherwise specified, we equip $\mathcal{M}(X)$, $\mathcal{P}(X)$, and $\mathcal{M}(X,g)$ with the weak* topology.

The space of real-valued (resp. complex-valued) Hölder continuous functions with an exponent $\alpha \in (0,1]$ on a compact metric space (X,d) is denoted as $C^{0,\alpha}(X,d)$ (resp. $C^{0,\alpha}((X,d),\mathbb{C})$). For each $\psi \in C^{0,\alpha}((X,d),\mathbb{C})$,

$$|\psi|_\alpha = \sup\left\{ \frac{|\psi(x)-\psi(y)|}{d(x,y)^\alpha} \,\middle|\, x,y \in X, x \neq y \right\}, \tag{0.1}$$

and the Hölder norm is defined as

$$\|\psi\|_{C^{0,\alpha}} = |\psi|_\alpha + \|\psi\|_\infty. \tag{0.2}$$

For given $f\colon X \to X$ and $\varphi \in C(X,\mathbb{C})$, we define

$$S_n\varphi(x) = \sum_{j=0}^{n-1} \varphi(f^j(x)) \tag{0.3}$$

and

$$W_n(x) = \frac{1}{n}\sum_{j=0}^{n-1} \delta_{f^j(x)} \tag{0.4}$$

for $x \in X$ and $n \in \mathbb{N}_0$. Note that when $n = 0$, by definition we always have $S_0\varphi = 0$, and by convention $W_0 = 0$.

Chapter 1
Introduction

Self-similar fractals have fascinated laymen and mathematicians alike due to their intrinsic beauty as well as mathematical sophistication. They appear naturally in mathematics and play important roles in the investigation of the corresponding areas of research. One particularly abundant source of self-similar fractals is the study of holomorphic dynamics, where they arise as Julia sets of rational functions and limit sets of Kleinian groups.

A powerful and fruitful point of view in the study of self-similar fractals is to look at them as metric spaces. On the other hand, due to their natural appearance in dynamics, self-similar fractals lie in the center of the interplay of dynamics and geometry. The investigation of metric and measure-theoretic properties of various self-similar fractals and their relation to dynamics and geometry has been actively carried out in different areas of mathematics.

Various tools in the study of general metric spaces become indispensable in the investigation of fractal spaces. Thanks to the new developments in the theory of quasiconformal geometry in recent years, more powerful tools become available and new perspectives become natural.

The classical theory of quasiconformal maps between Euclidean spaces dates back to the works of H. Grötzsch and L.V. Ahlfors in the early 20th century [Kü97, Ah82]. Since the groundbreaking work of O. Teichmüller on the classical moduli problem for Riemann surfaces around 1940 and later D.P. Sullivan's no-wandering-domain theorem in complex dynamics in the 1980s [Su85], nowadays the theory of planar quasiconformal maps is considered a standard tool in many areas of complex analysis such as Techmüller theory and holomorphic dynamics. Many such applications rely on an existence theorem for planar quasiconformal maps known as the Measurable Riemann Mapping Theorem. In higher dimensions, though, there is no counterpart for such an existence theorem. However, the importance of the theory of quasiconformal maps in higher dimensions became evident when G.D. Mostow used it in his celebrated rigidity theorems for rank-one symmetric spaces in the early

Z. Li, *Ergodic Theory of Expanding Thurston Maps*, Atlantis Studies in Dynamical Systems 4, DOI 10.2991/978-94-6239-174-1_1

1970s [Mos73]. This also inspired the generalization of the theory from $\mathbb{R}^n$ to metric spaces (see for example, [KR95, Pan89, HK98]).

We recall that in a metric space context, a homeomorphism $f: X \to Y$ between metric spaces (X, d_X) and (Y, d_Y) is *quasiconformal* if there exists a constant $H \geq 1$ such that

$$H_f(x) := \limsup_{r \to 0^+} \frac{\sup\{d_Y(f(x'), f(x)) \mid d_X(x', x) \leq r\}}{\inf\{d_Y(f(x'), f(x)) \mid d_X(x', x) \geq r\}} \leq H$$

for all $x \in X$. This definition is equivalent to the classical definition of a quasiconformal map in the context of Euclidean spaces, which we refer the reader to [Bon06]. In a context of a Euclidean space, it means, roughly speaking, that infinitesimal balls are mapped to infinitesimal ellipsoids with uniformly controlled eccentricity. In general the above definition is too weak to be useful.

A stronger and much more useful concept in the study of general metric spaces is the notion of a quasisymmetric map [TV80]. A homeomorphism $f: X \to Y$ is called *quasisymmetric* if there exists a homeomorphism $\eta: [0, +\infty) \to [0, +\infty)$ such that

$$\frac{d_Y(f(x), f(y))}{d_Y(f(x), f(z))} \leq \eta\left(\frac{d_X(x, y)}{d_X(x, z)}\right),$$

for all $x, y, z \in X$ with $x \neq z$. Roughly speaking, the above definition requires that balls to be mapped to "round" sets with quantitative control for their "eccentricity". This is a global version of the geometric property of a quasiconformal map. These two notions coincide in the context of Euclidean spaces.

The notion of quasisymmetry has been proved to be central in the study of various fractal metric spaces (see for example, [BM17, Bon11, BKM09]) and metric uniformization problems (see for example, [TV80, DS97, BonK02, Wi07]).

We now draw our attention back to the Riemann sphere.

Through the introduction of quasiconformal geometry techniques in his proof of the no-wandering-domain theorem in the 1980s [Su85], D.P. Sullivan revolutionized the field of complex dynamics. Originally, the theory of complex dynamics dates back to the work of G. Kœnigs, E. Schröder, and others in the 19th century. This subject, concentrating on the study of iterated rational maps on the Riemann sphere, was developed into an active and fascinating area of research, thanks to the remarkable works of S. Lattès, C. Carathéodory, P. Fatou, G. Julia, P. Koebe, L. Ahlfors, L. Bers, M. Herman, A. Douady, D.P. Sullivan, J.H. Hubbard, W.P. Thurston, J.C. Yoccoz, C. McMullen, J. Milnor, M. Lyubich, M. Shishikura, and many others. Modern research in complex dynamics centers at the study of fractals appearing in the dynamical space, namely the Julia sets, as well as ones in the parameter space such as the well-known Mandelbrot set.

In the early 1980s, D.P. Sullivan introduced a "dictionary", known as *Sullivan's dictionary* nowadays, linking the theory of complex dynamics with another classical area of conformal dynamical systems, namely, geometric group theory, mainly concerning the study of Kleinian groups acting on the Riemann sphere. Many dynamical objects in both areas can be similarly defined and results similarly proven, yet essential and important differences remain.

Sullivan's dictionary provides many connections and guiding intuitions between Kleinian groups and iterated rational maps on the Riemann sphere. A *Kleinian group* G is a discrete subgroup of the conformal automorphism group $\mathrm{Aut}\,(\widehat{\mathbb{C}})$ of the Riemann sphere $\widehat{\mathbb{C}}$ and a rational map is a quotient of two polynomails on $\widehat{\mathbb{C}}$. For both subjects, there are common themes in deformation theories [MS98], and combinatorial classification theories [Mc95, Pi03]. The geometric structures of the associated fractals in both subjects are also closely related [Mc98, Mc00, SU00, SU02]. For more detailed discussions of the correspondence between these two subjects, we refer the reader to [Mc95, Mc08, HP09] and references therein.

One natural question to ask when one investigates the essential features of these two subjects is the following: "*What is special about conformal dynamical systems in a wider class of dynamical systems characterized by suitable metric-topological conditions?*"

This general question has inspired much research in both subjects (see for example, [DH93, HSS09, ZJ09, CT11, CT15, BonK02, BonK05, KK00, BM17, HP09, HP14, Th16]). Often certain combinatorial information of the dynamical systems and the metric geometry of the associated fractal spaces play an important role in such investigations.

In geometric group theory, the above question is related to a well-known conjecture by J.W. Cannon [Ca94]. Recall that a Kleinian group G extends isometrically to the hyperbolic 3-space $\mathbb{H}^3$. Cannon's Conjecture predicts that for every Gromov hyperbolic group G whose boundary at infinity $\partial_\infty G$ is homeomorphic to the 2-sphere S^2, there should exist a discrete, cocompact, and isometric action of G on the hyperbolic 3-space $\mathbb{H}^3$. Here we can consider Gromov hyperbolic groups G with 2-sphere boundary $\partial_\infty G$ as metric-topological systems generalizing the conformal dynamical systems in this context, namely, certain Kleinian groups. Recall that there are natural metrics d_{vis} on $\partial_\infty G$ called *visual metrics*. These metrics are unique up to snowflake equivalence. From the point of view of metric properties, one can formulate Cannon's Conjecture in the following way: Let G be a Gromov hyperbolic group, then $\partial_\infty G$ is homeomorphic to the 2-sphere if and only if the metric space $(\partial_\infty, d_{\mathrm{vis}})$ is quasisymmetrically equivalent to the Riemann sphere $\widehat{\mathbb{C}}$. Note that two metric spaces are *quasisymmetrically equivalent* if there exists a quasisymmetric homeomorphism between them. Considerable amount of efforts have been made to establish Cannon's Conjecture, leading to various partial results (see for example, [BonK05, BouK13, Mar13]), but the conjecture still remains open.

Cannon's Conjecture translates via Sullivan's dictionary to the celebrated characterization theorem of rational maps in complex dynamics by W.P. Thurston [DH93]. In this context, the metric-topological dynamical systems that generalize postcritically-finite rational maps on the Riemann sphere are called *Thurston maps*. These are (non-homeomorphic) branched covering maps on the 2-sphere S^2 whose finitely many critical points are all preperiodic. Thurston's combinatorial characterization of rational maps asserts that a Thurston map is essentially a rational map if and only if there does not exist so-called *Thurston obstruction*, i.e., a collection of simple closed curves on S^2 subject to certain conditions [DH93].

By imposing some additional condition of expansion, thus restricting to a subclass of Thurston maps, a characterization theorem of rational maps from a metric space point of view has been established in this context by M. Bonk and D. Meyer [BM17], and P. Haïssinsky and K. Pilgrim [HP09]. Roughly speaking, we say that a Thurston map is expanding if for each pair of points $x, y \in S^2$, their preimages under iterations of the map get closer and closer. See Definition 2.10 for a precise formulation. We also refer to [BM17, Proposition 6.3] for a list of equivalent definitions. For each expanding Thurston map, we can equip the 2-sphere S^2 with a natural class of metrics d, called *visual metrics*, that are quasisymmetrically equivalent to each other. As the name suggests, these metrics are constructed in a similar fashion as the visual metrics on the boundary $\partial_\infty G$ of a Gromov hyperbolic group G (see [BM17, Chap. 8] for details, and see [HP09] for a related construction). In the language above, the following theorem was obtained in [BM17, HP09].

Theorem 1.1 (M. Bonk and D. Meyer, P. Haïssinsky and K. Pilgrim) *An expanding Thurston map is conjugate to a rational map if and only if the sphere (S^2, d) equipped with a visual metric d is quasisymmetrically equivalent to the Riemann sphere $\widehat{\mathbb{C}}$ equipped with the spherical metric.*

The dynamics induced by iterations of expanding Thurston maps mentioned above is going to be the main subject matter of this monograph.

Various characterization theorems of rational maps correspond to Cannon's Conjecture via Sullivan's dictionary. M. Bonk and B. Kleiner proved in [BonK05] a weak form of Cannon's Conjecture by adding an additional condition on the dimensions of the visual metrics. From the same metric property point of view, P. Haïssinsky and K. Pilgrim established in [HP14] a sufficient condition for an expanding Thurston map to be essentially a rational map. Their theorem asserts that if an Ahlfors regular metric d' that is quasisymmetrically equivalent to a visual metric d of an expanding Thurston map $f\colon S^2 \to S^2$ realizes the Ahlfors regular conformal dimension $\operatorname{confdim}_{\mathrm{AR}}(f)$ of f, then f is conjugate to a rational map except for special cases of so-called obstructed Lattès examples. Here the Ahlfors regular conformal dimension $\operatorname{confdim}_{\mathrm{AR}}(f)$ of f is defined as the infimum of the Hausdorff dimension of all Ahlfors regular metrics that are quasisymmetrically equivalent to a visual metric of f. A metric space (X, d) is *Ahlfors regular* of dimension Q provided there is a Radon measure μ and a constant $C > 1$ such that

$$\frac{1}{C} r^Q \leq \mu(B_d(x, r)) \leq C r^Q$$

for $x \in X$ and $r \in (0, \operatorname{diam}_d(X)]$.

Due to important and fruitful applications of Thurston's theorem, many authors have worked on extending it beyond postcritically-finite rational maps using similar combinatorial obstructions. See for example, J.H. Hubbard, D. Schleicher, M. Shishikura's work on some postcritically-finite exponential maps [HSS09]; G. Cui and L. Tan's and G. Zhang and Y. Jiang's works on hyperbolic rational maps [CT11, ZJ09]; G. Zhang's work on certain rational maps with Siegel disks [Zh08]; X. Wang's work on certain rational maps with Herman rings [Wan14]; and G. Cui and L. Tan's work on some geometrically finite rational maps [CT15]. The characterization theorems of rational maps from [BM17, HP09, HP14] mentioned above provide an entirely new perspective (from properties of metric spaces) to the classical combinatorial classification theorem of rational maps by W.P. Thurston and its various generalizations.

The conditions regarding the Ahlfors regular conformal dimension in [BonK05, HP14] also reveal the relevance of dimension theory in our metric and dynamical context.

The notions of fractal dimension are widely used in many different areas of mathematics and natural sciences nowadays. C. Carathéodory, F. Hausdorff, and A.S. Besicovich laid the foundation of dimension theory in the early twentieth century. The early investigation centered around the Hausdorff dimension, which serves as an appropriate notion to measure the complexity of topological and geometric structures of subsets in metric spaces that are similar to the well-known Cantor set. Thanks to the powerful tools of computer visualization, the study of fractal objects became popular in mathematics and natural sciences. The ideas of fractal dimension were explored extensively by practitioners in sciences and applied mathematics, usually heuristically, well before rigorous mathematical theories were developed.

In the study of dynamical systems, it is strongly believed that there is a deep connection between the topology and geometry of invariant fractal sets and properties of the dynamical system acting on them. For a discussion on the relationship between various notions of fractal dimension and invariants of the dynamical systems such as Lyapunov exponents and entropy, see for example, [GOY88].

Many methods and tools developed in the study of dynamical systems have been proved to be useful in the investigation of notions of fractal dimension. The thermodynamical formalism, and more generally, ergodic theory, are such important tools. For close relationship between thermodynamical formalism and fractal dimension theory in dynamical systems, see [Pe97, Barr11, PU10].

Ergodic theory has been an important tool in the study of dynamical systems in general. The investigation of the existence and uniqueness of invariant measures and their properties has been a central part of ergodic theory. However, a dynamical system may possess a large class of invariant measures, some of which may be more interesting than others. It is therefore crucial to examine the relevant invariant measures.

The *thermodynamical formalism* is one such mechanism to produce invariant measures with some nice properties under assumptions on the regularity of their *Jacobian functions*. More precisely, for a continuous transformation on a compact metric space, we can consider the *topological pressure* as a weighted version of the *topological entropy*, with the weight induced by a real-valued continuous function, called a *potential*. The Variational Principle identifies the topological pressure with the supremum of its measure-theoretic counterpart, the *measure-theoretic pressure*, over all invariant Borel probability measures [Bow75, Wal76]. Under additional regularity assumptions on the transformation and the potential, one gets existence and uniqueness of an invariant Borel probability measure maximizing the measure-theoretic pressure, called the *equilibrium state* for the given transformation and the potential. Often the Jacobian function for the transformation with respect to the equilibrium state is prescribed by a function induced by the potential. The study of the existence and uniqueness of the equilibrium states and their various properties such as ergodic properties, equidistribution, fractal dimensions, etc., has been the main motivation for much research in the area.

This theory, as a successful approach to choosing relevant invariant measures, was inspired by statistical mechanics, and created by D. Ruelle, Ya. Sinai, and others in the early seventies [Dob68, Si72, Bow75, Wal82]. Since then, the thermodynamical formalism has been applied in many classical contexts (see for example, [Bow75, Ru89, Pr90, KH95, Zi96, MauU03, BS03, Ol03, Yu03, PU10, MayU10]). However, beyond several classical dynamical systems, even the existence of equilibrium states is largely unknown, and for those dynamical systems that do possess equilibrium states, often the uniqueness is unknown or at least requires additional conditions. The investigation of different dynamical systems from this perspective has been an active area of current research.

This monograph is intended as an introduction to the ergodic theory of expanding Thurston maps. More specifically, it focuses on the properties of important invariant measures such as the measure of maximal entropy and more generally, equilibrium states corresponding to Hölder continuous potentials, and their relationship with periodic points and preimage points.

We consent ourselves in this monograph by providing a foundation and a model case for more involved studies in this area on more general branched covering maps on S^2 such as ones investigated in [ZJ09, CT11, CT15], or between certain general topological spaces such as coarse expanding conformal maps from [HP09]; or more general and probably more useful potentials with logarithmic singularities similar to ones in [MayU10].

We develop the ergodic theory for expanding Thurston maps in three stages. In the first stage, we investigate various properties of the measure of maximal entropy (see Sect. 3.2 for definition) by direct and elementary arguments using the explicit combinatorial and geometric information of the maps. Among other things, we obtain very strong equidistribution results for preimage points, periodic points, and preperiodic points with respect to the measure of maximal entropy (see Theorems 4.2, 4.3, and Corollary 4.4). In order to establish the existence, uniqueness, and various other properties of equilibrium states for Hölder continuous potentials, one needs to apply more powerful tools, namely, the theory of thermodynamical formalism. This is what we do in the second stage. The equidistribution results with respect to equilibrium states we get from thermodynamical formalism are for preimage points only, and have less flexible choice of weight at each point compared to the corresponding results we get in the first stage (see Theorem 5.1). In order to get equidistribution results with respect to equilibrium states for periodic points, we apply in the last stage another machinery due to Y. Kifer [Ki90], which leads to some large deviations principles (see Theorem 7.1) which are stronger than equidistribution results. We are able to use this machinery for a subclass of our maps, more precisely, expanding Thurton maps without periodic critical points. For these maps, we establish the upper semi-continuity of the measure-theoretic entropy function by investigating certain weak expansion properties of our dynamical systems. However, upper semi-continuity of the measure-theoretic entropy and equidistribution of periodic points with respect to equilibrium states still remain open for expanding Thurston maps with a periodic critical point.

We now discuss our approaches in more details.

Arguably the most important measure for a dynamical system is its *measure of maximal entropy*. By definition, it is an invariant Borel probability measure that maximizes the measure-theoretic entropy. Thanks to the pioneering work of R. Bowen, D. Ruelle, P. Walters, Ya. Sinai, M. Lyubich, R. Mañé, and many others, existence and uniqueness results for the measure of maximal entropy are known for uniformly expansive continuous dynamical systems, distance expanding continuous dynamical systems, uniformly hyperbolic smooth dynamical systems, and rational maps on the Riemann sphere. In many cases, the measure of maximal entropy is also the asymptotic distribution of the period points (see [Par64, Si72, Bow75, Ly83, FLM83, Ru89, PU10]).

Expanding Thurston maps do not fall into any class of the classical dynamical systems mentioned above (see Chap. 6 for a more detailed discussion). So we have to first investigate the existence and uniqueness of such measures. As a consequence of their general results in [HP09], P. Haïssinsky and K. Pilgrim proved that for each expanding Thurston map, there exists a measure of maximal entropy and that the measure of maximal entropy is unique for an expanding Thurston map without periodic critical points. M. Bonk and D. Meyer then proved the existence and uniqueness of the measure of maximal entropy for all expanding Thurston maps using an explicit combinatorial construction [BM17]. Some equidistribution results for periodic critical points and iterated preimages with respect to the measure of maximal entropy were obtained in [HP09]. Following the philosophy of M. Bonk and D. Meyer, we establish in Chap. 4 stronger equidistribution results for (pre)periodic points and iterated preimages with respect to the measure of maximal entropy in our context. In order to do so, we carefully investigate the locations of fixed points in relation to the Markov partitions. We also establish an exact formula for the number of fixed points for an expanding Thurston map (see Theorem 4.1), which is analogous to the corresponding formula for rational maps (see for example, [Mil06, Theorem 12.1]).

After all, the measure of maximal entropy is just one important invariant measure. In order to investigate a larger class of important invariant measures, one needs to apply more powerful tools from thermodynamical formalism.

We establish the existence and uniqueness of the equilibrium state, denoted by μ_ϕ, for a Hölder continuous potential $\phi\colon S^2 \to \mathbb{R}$. Here S^2 is equipped with a visual metric. This generalizes the existence and uniqueness of the measure of maximal entropy of an expanding Thurston map in [HP09, BM17]. We also prove that the measure-preserving transformation f of the probability space (S^2, μ_ϕ) is *exact* (see Definition 5.40), and in particular, mixing and ergodic (Theorem 5.41 and

Corollary 5.44). This generalizes the corresponding results in [BM17, HP09] for the measure of maximal entropy to our context.

We quickly review some key concepts. For an expanding Thurston map $f\colon S^2 \to S^2$ and a continuous function $\psi\colon S^2 \to \mathbb{R}$, and each f-invariant Borel probability measure μ on S^2, we have an associated quantity,

$$P_\mu(f, \psi) = h_\mu(f) + \int \psi \, \mathrm{d}\mu,$$

called the *measure-theoretic pressure* of f for μ and ψ, where $h_\mu(f)$ is the measure-theoretic entropy of f for μ. The well-known Variational Principle (see for example, [PU10, Theorem 3.4.1]) asserts that

$$P(f, \psi) = \sup P_\mu(f, \psi), \tag{1.1}$$

where the supremum is taken over all f-invariant Borel probability measures μ, and $P(f, \psi)$ is the *topological pressure* of f with respect to ψ defined in (3.1). A measure μ that attains the supremum in (1.1) is called an *equilibrium state* for f and ψ.

We assume for now that ψ is Hölder continuous (with respect to a given visual metric for f on S^2). One characterization of the topological pressure in our context is given by the following formula (Proposition 5.19):

$$P(f, \psi) = \lim_{n \to +\infty} \frac{1}{n} \log \sum_{y \in f^{-n}(x)} \deg_{f^n}(y) \exp(S_n\psi(y)), \tag{1.2}$$

for each $x \in S^2$, independent of x, where $\deg_{f^n}(y)$ is the local degree of f^n at y and $S_n\psi(y) = \sum_{i=0}^{n-1} \psi(f^i(y))$.

An important tool that we use to find the equilibrium state and to establish its uniqueness, is the *Ruelle operator* $\mathcal{L}_\psi$ on the Banach space $C(S^2)$ of real-valued continuous functions on S^2, given by

$$\mathcal{L}_\psi(u)(x) = \sum_{y \in f^{-1}(x)} \deg_f(y) u(y) \exp(\psi(y)),$$

for $u \in C(S^2)$ and $x \in S^2$.

The Ruelle operator plays a central role in the thermodynamical formalism, and has been studied carefully for various dynamical systems (see for example, [Bow75, Ru89, Pr90, Zi96, MauU03, PU10, MayU10]). Some of the ideas that we apply in Chap. 5 for its investigation are well-known and repeatedly used in the literature, see for example [PU10, Zi96].

A main difficulty of our analysis comes from the lack of uniform expansion property that arises from the existence of critical points (i.e., branch points of a branched covering map). As an example, identities of the form (1.2) that are usually easy to derive for classical dynamical systems (see for example, [PU10, Proposition 4.4.3]) become difficult to verify directly in our context.

The main results that we obtain via the thermodynamical formalism in Chap. 5 is summarized in Theorem 5.1. A rational Thurston map is expanding if and only if it has no periodic critical points (see [BM17, Proposition 2.3]). We also get a version of Theorem 5.1 in the rational case (see Corollary 5.2).

We remark here that the existence and uniqueness of the equilibrium state for a *general* rational map R on the Riemann sphere and a real-valued Hölder continuous potential ϕ can be established under the additional assumption that $\sup\{\phi(z) \mid z \in J(R)\} < P(R,\phi)\}$, where $J(R)$ is the Julia set of R and $P(R,\phi)$ is the topological pressure of R with respect to ϕ (see [DU91, Pr90, DPU96]). This assumption can sometimes be dropped: one can either restrict to certain subclasses of rational maps, such as topological Collet-Eckmann maps, see [CRL11], or hyperbolic rational maps (more generally, distance-expanding maps), see [PU10]; or one can impose other conditions on the function ϕ, such as hyperbolicity of ϕ, see [IRRL12]. It is easy to check that a rational expanding Thurston map is topological Collet-Eckmann.

As a consequence of the proof of the uniqueness of the equilibrium states, we also obtain equidistribution results (5.1) and (5.2) for the iterated preimages with respect to the equilibrium states as stated in Theorem 5.1. However, similar results for periodic points are inaccessible by the usual techniques from thermodynamical formalism due to technical difficulties arising from the existence of critical points.

Rather than trying to establish the equidistribution results for periodic points directly, we derive in Chap. 7 some stronger results using a general framework of Y. Kifer [Ki90]. More precisely, we obtain *level-2 large deviation principles* for periodic points (as well as iterated preimages) with respect to equilibrium states in the context of expanding Thurston maps without periodic critical points and Hölder continuous potentials. We use a variant of Y. Kifer's result formulated by H. Comman and J. Rivera-Letelier [CRL11], which is recorded in Theorem 7.7 for the convenience of the reader. For related results on large deviation principles in the context of rational maps on the Riemann sphere under additional assumptions, see [PSh96, PSr07, XF07, PRL11, Com09, CRL11].

Let us discuss these results in more details. We denote the space of Borel probability measures on a compact metric space X equipped with the weak* topology by $\mathcal{P}(X)$. A sequence $\{\Omega_n\}_{n\in\mathbb{N}}$ of Borel probability measures on $\mathcal{P}(X)$ is said to satisfy a *level-2 large deviation principle with rate function* $I\colon \mathcal{P}(X) \to [0,+\infty]$ if for each closed subset $\mathfrak{F}$ of $\mathcal{P}(X)$ and each open subset $\mathfrak{G}$ of $\mathcal{P}(X)$ we have

$$\limsup_{n\to+\infty} \frac{1}{n} \log \Omega_n(\mathfrak{F}) \leq -\inf\{I(x) \mid x \in \mathfrak{F}\},$$

and

$$\liminf_{n\to+\infty} \frac{1}{n} \log \Omega_n(\mathfrak{G}) \geq -\inf\{I(x) \mid x \in \mathfrak{G}\}.$$

See Sect. 7.1 for more details. We also refer the reader to [CRL11, Sect. 2.5] and the references therein for a more systematic introduction to the theory of large deviation principles.

In order to apply Theorem 7.7, we just need to verify three conditions:

(1) The existence and uniqueness of the equilibrium state.
(2) Some characterization of the topological pressure (such as Propositions 7.8 and 7.9).
(3) The upper semi-continuity of the measure-theoretic entropy.

The first condition is established by Theorem 5.1. The second condition is weaker than the equidistribution results, and is within reach. The last condition seems to be difficult to verify directly.

In order to establish the upper semi-continuity of the measure-theoretic entropy, we need to take a closer look at the expansion property of our dynamical systems.

In the study of discrete-time dynamical systems, various conditions can be imposed upon the map to simplify the orbit structures, which in turn lead to results about the dynamical system under consideration. One such well-known condition is expansiveness. Roughly speaking, a map is expansive if no two distinct orbits stay close forever. Expansiveness plays an important role in the investigation of hyperbolicity in smooth dynamical systems, and in complex dynamics in particular (see for example, [Man87, PU10]).

In the context of continuous maps on compact metric spaces, there are two weaker notions of expansion, called *h-expansiveness* and *asymptotic h-expansiveness*, introduced by R. Bowen [Bow72] and M. Misiurewicz [Mis73], respectively. Forward-expansiveness implies h-expansiveness, which in turn implies asymptotic h-expansiveness [Mis76]. Both of these weak notions of expansion play important roles in the study of smooth dynamical systems (see [Burg11, DFPV12, DM09, DN05, LVY13]). Moreover, any smooth map on a compact Riemannian manifold is asymptotically h-expansive [Buz97]. Recently, N.-P. Chung and G. Zhang extended these concepts to the context of a continuous action of a countable discrete sofic group on a compact metric space [CZ15].

M. Misiurewicz showed that asymptotic h-expansiveness guarantees that the measure-theoretic entropy $\mu \mapsto h_\mu(f)$ is upper semi-continuous [Mis76].

To be a bit more precisely, we let (X, d) be a compact metric space, and $g: X \to X$ a continuous map on X. Denote, for $\varepsilon > 0$ and $x \in X$,

$$\Phi_\varepsilon(x) = \{y \in X \mid d(g^n(x), g^n(y)) \leq \varepsilon \text{ for all } n \geq 0\}.$$

The map g is called *forward expansive* if there exists $\varepsilon > 0$ such that $\Phi_\varepsilon(x) = \{x\}$ for all $x \in X$. By R. Bowen's definition in [Bow72], the map g is *h-expansive* if there exists $\varepsilon > 0$ such that the topological entropy $h_{\text{top}}(g|_{\Phi_\varepsilon(x)}) = h_{\text{top}}(g, \Phi_\epsilon(x))$ of g restricted to $\Phi_\varepsilon(x)$ is 0 for all $x \in X$. One can also formulate asymptotic h-expansiveness in a similar spirit, see for example, [Mis76, Sect. 2]. However, in this monograph, we will adopt equivalent formulations from [Dow11]. See Sect. 3.4 for details.

Another way to formulate forward expansiveness is via distance expansion. We say that $g\colon X \to X$ is *distance-expanding* (with respect to the metric d) if there exist constants $\lambda > 1$, $\eta > 0$, and $n \in \mathbb{N}$ such that for all $x, y \in X$ with $d(x, y) \leq \eta$, we have $d(g^n(x), g^n(y)) \geq \lambda d(x, y)$. If g is forward expansive, then there exists a metric ρ on X such that the metrics d and ρ induce the same topology on X and g is distance-expanding with respect to ρ (see for example, [PU10, Theorem 4.6.1]). Conversely, if g is distance-expanding, then it is forward expansive (see for example, [PU10, Theorem 4.1.1]). So roughly speaking, if g is forward expansive, then the distance between two points that are close enough grows exponentially under forward iterations of g.

Since a Thurston map, by definition, has to be a branched covering map, we can always find two distinct points that are arbitrarily close to a critical point (thus arbitrarily close to each other) and that are mapped to the same point. Thus a Thurston map cannot be forward expansive. The expansion property of expanding Thurston maps may nevertheless still seem to be quite strong. However, as a part of the main theorem for Chap. 6, we will show that no expanding Thurston map is h-expansive. Moreover, we prove that an expanding Thurston map is asymptotically h-expansive if and only if it has no periodic critical points (see Theorem 6.1).

When R. Bowen introduced h-expansiveness in [Bow72], he mentioned that no diffeomorphism of a compact manifold was known to be not h-expansive. M. Misiurewicz then produced an example of a diffeomorphism that is not asymptotically h-expansive [Mis73]. M. Lyubich showed that each rational map is asymptotically h-expansive [Ly83]. J. Buzzi established asymptotic h-expansiveness of any C^∞-map on a compact Riemannian manifold [Buz97]. Examples of C^∞-maps that are not h-expansive were given by M.J. Pacifico and J.L. Vieitez [PV08]. Our Theorem 6.1 implies that no rational expanding Thurston map (i.e., any postcritically-finite rational map whose Julia set is the whole sphere (see [BM17, Proposition 2.3]) is h-expansive.

Expanding Thurston maps may be the first example of a class of a priori non-differentiable maps that are not h-expansive but may be asymptotically h-expansive depending on some properties of orbits of critical points.

As an immediate consequence of Theorem 6.1 and the result of J. Buzzi [Buz97] mentioned above, we get that an expanding Thurston map with at least one periodic critical point cannot be conjugate to a C^∞-map (see Corollary 6.2). It partially answers a question of K. Pilgrim (see Problem 2 in [BM10, Sect. 21]). See Remark 6.3 for a sketch of an elementary proof of this fact.

Returning back to our original motivation from equidistribution results and large deviation principles, we get the upper semi-continuity of the measure-theoretic entropy function for expanding Thurston maps without periodic critical points (see Corollary 6.4) as a consequence of Theorem 6.1 and M. Misiurewicz's result in [Mis76] mentioned above. As an immediate consequence of the upper semi-continuity of the measure-theoretic entropy, we obtain the existence of equilibrium states for such Thurston maps and all continuous potentials (see Theorem 6.5).

With the upper semi-continuity of the measure-theoretic entropy function, we are able to apply the machinery of Y. Kifer reformulated by H. Common and

J. Rivera-Letelier mentioned above to get level-2 large deviation principles for expanding Thurston maps with no periodic critical points in Chap. 7 (see Theorem 7.7). As a consequence, for such maps we get the equidistribution of periodic points and preimage points with more flexible choices of weights (see Corollary 7.3), which was our motivation to consider the weak expansion properties of expanding Thurston maps in Chap. 6. For a discussion of extending such results to maps with at least one periodic critical point and related open questions, we refer the reader to the beginning of Chap. 7.

We will now give a brief description of the structure of this monograph.

Chapter 2 provides an introduction to Thurston maps. We first define branched covering maps on S^2 and Thurston maps in Sect. 2.2. We then introduce cell decompositions $\mathbf{D}^n(f, \mathscr{C})$, $n \in \mathbb{N}$, of S^2 induced by a Thurston map $f: S^2 \to S^2$ and a Jordan curve $\mathscr{C} \subseteq S^2$ containing the postcritical points post f in Sect. 2.3. We then define expanding Thurston maps and combinatorially expanding Thurston maps in Sect. 2.4. Next, we discuss visual metrics on S^2 for an expanding Thurston map in Sect. 2.5. We summarize properties of visual metrics from [BM17], especially the relation between visual metrics and the cell decompositions. We establish a few properties of expanding Thurston maps in this chapter that will be used later.

In Chap. 3, we review some key concepts of ergodic theory. We first recall the usual notions of covers and partitions. Next, measure-theoretic entropy and topological entropy, as well as measure-theoretic pressure and topological pressure are introduced before measures of maximal entropy and equilibrium states are defined. We then formulate the Ruelle operator $\mathscr{L}_\psi$ for an expanding Thurston map and a complex-valued continuous function ψ on S^2 in Sect. 3.3. We argue that it is well-defined on the space of complex-valued continuous functions on S^2 and discuss some of its properties. We end this chapter by reviewing the notions of topological conditional entropy and topological tail entropy, and recalling the definitions of h-expansiveness and asymptotic h-expansiveness using these notions. We adopt the terminology and formulations by T. Downarowicz in [Dow11]. Chapter 4 is devoted to the study of periodic points and the measure of maximal entropy for an expanding Thurston map. In Sect. 4.1, we investigate the fixed points, periodic points, and preperiodic points of the expanding Thurston maps. In particular, we study the location of periodic points and establish a formula for the number of fixed points of each expanding Thurston map (Theorem 4.1). In Sect. 4.2, we prove a number of equidistribution results for periodic points and iterated preimages with respect to the measure of maximal entropy using the exact combinatorial information we obtained in Sect. 4.1. In Sect. 4.3, we show that for each expanding Thurston map f with its measure of maximal entropy μ_f, the measure-preserving dynamical system (S^2, f, μ_f) is a factor, in the category of measure-preserving dynamical systems, of the measure-preserving dynamical system of the left-shift operator on the one-sided infinite sequences of $\deg f$ symbols together with its measure of maximal entropy.

In Chap. 5, we investigate the existence, uniqueness, and other properties of equilibrium states for an expanding Thurston map. The main tool for this chapter is the thermodynamical formalism. In Sect. 5.1, we state the assumptions on some of the objects in the remaining part of this monograph, which we are going to repeatedly refer to later as *the Assumptions*. We emphasize that not all assumptions are assumed in all the statements in the subsequent chapters, and that in fact we have to gradually remove the dependence on some of the assumptions before proving our main results. In Sect. 5.2, following the ideas from [PU10, Zi96], we use the thermodynamical formalism to prove the existence of equilibrium states for an expanding Thurston map $f: S^2 \to S^2$ and a real-valued Hölder continuous potential $\phi: S^2 \to \mathbb{R}$. In Sect. 5.3, we establish the uniqueness of the equilibrium state μ_ϕ in this context. We use the idea in [PU10] to apply the Gâteaux differentiability of the topological pressure function and some techniques from functional analysis. In Sect. 5.4, we prove that the measure-preserving transformation f of the probability space (S^2, μ_ϕ) is exact (Theorem 5.41), where the equilibrium state μ_ϕ is non-atomic (Corollary 5.42). It follows in particular that the transformation f is mixing and ergodic (Corollary 5.44). We show in Sect. 5.5 that if ϕ and ψ are two real-valued Hölder continuous functions with the corresponding equilibrium states μ_ϕ and μ_ψ, respectively, then $\mu_\phi = \mu_\psi$ if and only if there exists a constant $K \in \mathbb{R}$ such that $\phi - \psi$ and $K \mathbb{1}_{S^2}$ are *co-homologous* in the space $C(S^2)$ of real-valued continuous functions, i.e., $\phi - \psi - K\mathbb{1}_{S^2} = u \circ f - u$ for some $u \in C(S^2)$ (Theorem 5.45). In Sect. 5.6, we first establish versions of equidistribution of iterated preimages with respect to the equilibrium state (Proposition 5.54), using results we obtain in Sect. 5.3. At the end of this chapter, following the idea of J. Hawkins and M. Taylor [HT03], we show that the equilibrium state μ_ϕ is almost surely the limit of

$$\frac{1}{n}\sum_{i=0}^{n-1} \delta_{q_i}$$

as $n \longrightarrow +\infty$ in the weak* topology, where q_0 is an arbitrary fixed point in S^2, and for each $i \in \mathbb{N}_0$, the point q_{i+1} is randomly chosen from the set $f^{-1}(q_i)$ with the probability of each $x \in f^{-1}(q_i)$ being q_{i+1} conditional on q_i proportional to the local degree of f at x times $\exp\big(\widetilde{\phi}(x)\big)$ (Theorem 5.55).

Chapter 6 focuses on the investigation of the weak expansion properties of expanding Thurston maps and the proof of Theorem 6.1, which asserts that an expanding Thurston map is asymptotically h-expansive if and only if f has no periodic critical points, and moreover, it is never h-expansive.

Chapter 7 is devoted to the study of large deviation principles and equidistribution results for periodic points and iterated preimages of expanding Thurston maps without periodic critical points. The idea is to apply a general framework devised by Y. Kifer [Ki90] to obtain level-2 large deviation principles, and to derive the equidistribution results as consequences. In Sect. 7.1, we give a brief introduction to level-2 large deviation principles in our context. We then establish some characterization of topological pressure in our context in Sect. 7.2 before providing a proof of Theorem 7.1, establishing level-2 large deviation principles in the context of expanding Thurston maps without periodic critical points and given Hölder continuous potentials. Finally in Sect. 7.4, we derive equidistribution of periodic points and iterated preimages with respect to the equilibrium state in our context with a fairly flexible choice of weight at each point.

Chapter 2
Thurston Maps

In this chapter, we introduce the dynamical systems that we are going to study, namely, expanding Thurston maps. We first recall briefly some history in Sect. 2.1, where we by no means intend to give a complete account of the development of the subject. We then introduce Thurston maps in Sect. 2.2 and certain cell decompositions of the 2-sphere S^2 induced by Thurston maps in Sect. 2.3. Next, we discuss various notions of expansion in our context and define expanding Thurston maps in Sect. 2.4. Two most important tools in the study of expanding Thurston maps are explored in the last two sections. The first tool is a natural class of metrics on the S^2, called visual metrics, discussed in Sect. 2.5. The second tool, discussed in Sect. 2.6, is the existence and properties of certain forward invariant Jordan curves on S^2, which induce nice partitions of the sphere. It is the geometric and combinatorial information we get from these tools that enables us to investigate the dynamical properties of expanding Thurston maps.

We prove in Lemma 2.12 that the union of all iterated preimages of an arbitrary point $p \in S^2$ of an expanding Thurston map is dense in S^2. We also summarize properties of visual metrics from [BM17], especially the relation between visual metrics and the cell decompositions, in Lemma 2.13 and the discussion that follows it. The fact that an expanding Thurston map is Lipschitz with respect to a visual metric is established in Lemma 2.15. M. Bonk and D. Meyer proved that for each expanding Thurston map f, there exists an f^n-invariant Jordan curve containing post f for each sufficiently large $n \in \mathbb{N}$ depending on f (see Theorem 2.16). We prove in Lemma 2.17 a slightly stronger version of this result, which carries additional combinatorial information of the Jordan curve. This lemma will be used in Chaps. 4 and 6. Finally, in Lemma 2.19, we prove that an expanding Thurston map locally expands the distance, with respect to a visual metric, between two points exponentially as long as they belong to one set in some particular partition of S^2 induced by a backward iteration of some Jordan curve on S^2. This observation, generalizing a result of M. Bonk and D. Meyer [BM17, Lemma 15.25], enables us to establish the distortion lemmas (Lemmas 5.3 and 5.4) in Sect. 5.2, which serve as the cornerstones for the mechanism of thermodynamical formalism that is essential in Chap. 5.

Z. Li, *Ergodic Theory of Expanding Thurston Maps*, Atlantis Studies in Dynamical Systems 4, DOI 10.2991/978-94-6239-174-1_2

2.1 Historical Background

The study of Thurston maps dates back to W.P. Thurston's celebrated combinatorial characterization theorem of postcritically-finite rational maps on the Riemann sphere among a class of more general continuous maps [DH93]. We call this class of continuous maps *Thurston maps* nowadays. Thurston's theorem asserts that a Thurston map is essentially a rational map if and only if there does not exist a so-called *Thurston obstruction*, i.e., a collection of simple closed curves on S^2 subject to certain conditions [DH93]. Due to the important and fruitful applications of Thurston's theorem, many authors have worked on extending it beyond postcritically-finite rational maps using similar combinatorial obstructions. See for example, J.H. Hubbard, D. Schleicher, M. Shishikura's work on some postcritically-finite exponential maps [HSS09]; G. Cui and L. Tan's and G. Zhang and Y. Jiang's works on hyperbolic rational maps [CT11, ZJ09]; G. Zhang's work on certain rational maps with Siegel disks [Zh08]; X. Wang's work on certain rational maps with Herman rings [Wan14]; and G. Cui and L. Tan's work on some geometrically finite rational maps [CT15].

It has since been a central theme in the study of conformal dynamical systems to search for Thurston-type theorems, i.e., characerizations of conformal dynamical systems in a wider class of dynamical systems satisfying suitable metric-topological conditions. See also [Th16, KPT15] for some remarkable recent works in this direction.

It is natural to ask for Thurston-type theorems from different points of view. One promising approach is from a point of view of metric space properties. In order to gain more precise metric estimates, groups of authors, notably M. Bonk and D. Meyer [BM17], P. Haïssinsky and K. Pilgrim [HP09], and J.W. Cannon, W.J. Floyd, and R. Parry [CFP07] started to impose natural notions of expansion in their respective contexts. These notions turned out to coincide in the context of expanding Thurston maps (see Sect. 2.4 for more details).

The existence of certain invariant Jordan curves as stated in Theorem 2.16 serves as foundation and starting point of the investigation of expanding Thurston maps. The special case of Theorem 2.16 for rational expanding Thurston maps was announced by M. Bonk during an Invited Address at the AMS Meeting at Athens, Ohio, in March 2004, where W.J. Floyd also mentioned a related result independently obtained by J.W. Cannon, W.J. Floyd, and R. Parry [CFP07]. Finally a Thurston-type theorem from a metric space point of view was obtained independently by M. Bonk and D. Meyer [BM17], and P. Haïssinsky and K. Pilgrim [HP09] in their respective contexts. See Theorem 1.1 in the case of expanding Thurston maps. Special cases of Theorem 1.1 go back to [Me02] and unpublished joint work of M. Bonk and B. Kleiner (see [BM17, Preface]).

2.2 Definition for Thurston Maps

Let S^2 denote an oriented topological 2-sphere. We first define branched covering maps on S^2. For detailed treatment of branched covering maps, we refer the reader to [BM17, Appendix 6].

A continuous map $f\colon S^2 \to S^2$ is called a *branched covering map* on S^2 if for each point $x \in S^2$, there exist a positive integer $d \in \mathbb{N}$, open neighborhoods U of x and V of $y = f(x)$, open neighborhoods U' and V' of 0 in $\widehat{\mathbb{C}}$, and orientation-preserving homeomorphisms $\varphi\colon U \to U'$ and $\eta\colon V \to V'$ such that $\varphi(x) = 0$, $\eta(y) = 0$, and

$$(\eta \circ f \circ \varphi^{-1})(z) = z^d$$

for each $z \in U'$. The above relation can be seen from the following diagram:

$$\begin{array}{ccc} x \in U & \xrightarrow{\ f\ } & y \in V \\ \Big\downarrow{\scriptstyle\varphi} & & \Big\downarrow{\scriptstyle\eta} \\ 0 \in U' & \xrightarrow[z \mapsto z^d]{} & 0 \in V'. \end{array}$$

The positive integer d above is uniquely determined by f and x, and is called the *local degree* of f at x, denoted by $\deg_f(x)$. The *(topological) degree* $\deg f$ of f can be calculated by

$$\deg f = \sum_{x \in f^{-1}(y)} \deg_f(x) \tag{2.1}$$

for $y \in S^2$ and is independent of y. If $f\colon S^2 \to S^2$ and $g\colon S^2 \to S^2$ are two branched covering maps on S^2, then so is $f \circ g$ (see [BM17, Lemma A.14]), and

$$\deg_{f\circ g}(x) = \deg_g(x)\deg_f(g(x)), \qquad \text{for each } x \in S^2, \tag{2.2}$$

and moreover,

$$\deg(f \circ g) = (\deg f)(\deg g). \tag{2.3}$$

A point $x \in S^2$ is a *critical point* of f if $\deg_f(x) \geq 2$. The set of critical points of f is denoted by $\operatorname{crit} f$. A point $y \in S^2$ is a *postcritical point* of f if $y = f^n(x)$ for some $x \in \operatorname{crit} f$ and $n \in \mathbb{N}$. The set of postcritical points of f is denoted by $\operatorname{post} f$. Note that $\operatorname{post} f = \operatorname{post} f^n$ for all $n \in \mathbb{N}$.

Definition 2.1 (*Thurston maps*) A Thurston map is a branched covering map $f\colon S^2 \to S^2$ on S^2 with $\deg f \geq 2$ and $\operatorname{card}(\operatorname{post} f) < +\infty$.

Example 2.2 We take two congruent Euclidean equilateral triangles ΔABC and $\Delta A'B'C'$, and then paste them along the boundary with A and A', B and B', and C

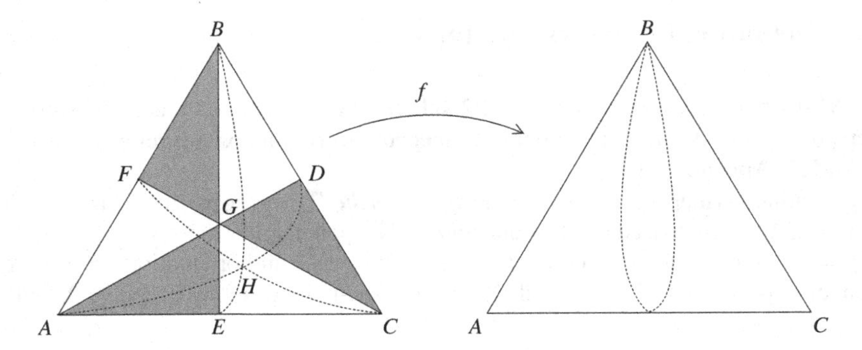

Fig. 2.1 A Thurston map from the barycentric subdivisions

and C' identified as shown in Fig. 2.1. We define a piecewise linear map $f: S^2 \to S^2$ by using the barycentric subdivision of each triangle. Firstly, the triangle ΔAGE is mapped linearly to the triangle ΔABC on the front of the sphere S^2 with $f(A) = A$, $f(E) = C$, and $f(G) = B$. We can then define f on ΔEGC in such a way that f maps ΔEGC linearly to the triangle $\Delta C'B'A'$ on the back of the sphere with $f(E) = C' = C$, $f(G) = B' = B$, and $f(C) = A' = A$. Similarly, we can extend f continuously to the whole sphere. We shade the preimages of the triangle ΔABC on the front of the sphere that are facing us as shown in Fig. 2.1.

It is easy to see that $\operatorname{crit} f = \{A, B, C, D, E, F, G, H\}$, and $\operatorname{post} f = \{A, B, C\}$. It is then clear that f is a Thurston map. We leave it as an exercise to show that f is an expanding Thurston map.

2.3 Cell Decompositions

We now recall the notion of cell decompositions of S^2. A *cell of dimension n* in S^2, $n \in \{1, 2\}$, is a subset $c \subseteq S^2$ that is homeomorphic to the closed unit ball $\overline{\mathbb{B}^n}$ in $\mathbb{R}^n$. We define the *boundary of* c, denoted by ∂c, to be the set of points corresponding to $\partial \mathbb{B}^n$ under such a homeomorphism between c and $\overline{\mathbb{B}^n}$. The *interior of* c is defined to be $\operatorname{inte}(c) = c \setminus \partial c$. For each point $x \in S^2$, the set $\{x\}$ is considered a *cell of dimension* 0 in S^2. For a cell c of dimension 0, we adopt the convention that $\partial c = \emptyset$ and $\operatorname{inte}(c) = c$.

We record the following three definitions from [BM17].

Definition 2.3 (*Cell decompositions*) Let $\mathbf{D}$ be a collection of cells in S^2. We say that $\mathbf{D}$ is a *cell decomposition of* S^2 if the following conditions are satisfied:

(i) The union of all cells in $\mathbf{D}$ is equal to S^2.
(ii) If $c \in \mathbf{D}$, then ∂c is a union of cells in $\mathbf{D}$.

(iii) For $c_1, c_2 \in \mathbf{D}$ with $c_1 \neq c_1$, we have $\operatorname{inte}(c_1) \cap \operatorname{inte}(c_2) = \emptyset$.
(iv) Every point in S^2 has a neighborhood that meets only finitely many cells in $\mathbf{D}$.

Definition 2.4 (*Refinements*) Let $\mathbf{D}'$ and $\mathbf{D}$ be two cell decompositions of S^2. We say that $\mathbf{D}'$ is a *refinement* of $\mathbf{D}$ if the following conditions are satisfied:

(i) Every cell $c \in \mathbf{D}$ is the union of all cells $c' \in \mathbf{D}'$ with $c' \subseteq c$.
(ii) For every cell $c' \in \mathbf{D}'$ there exits a cell $c \in \mathbf{D}$ with $c' \subseteq c$.

Definition 2.5 (*Cellular maps and cellular Markov partitions*) Let $\mathbf{D}'$ and $\mathbf{D}$ be two cell decompositions of S^2. We say that a continuous map $f : S^2 \to S^2$ is *cellular* for $(\mathbf{D}', \mathbf{D})$ if for every cell $c \in \mathbf{D}'$, the restriction $f|_c$ of f to c is a homeomorphism of c onto a cell in $\mathbf{D}$. We say that $(\mathbf{D}', \mathbf{D})$ is a *cellular Markov partition* for f if f is cellular for $(\mathbf{D}', \mathbf{D})$ and $\mathbf{D}'$ is a refinement of $\mathbf{D}$.

Let $f : S^2 \to S^2$ be a Thurston map, and $\mathscr{C} \subseteq S^2$ be a Jordan curve containing $\operatorname{post} f$. Then the pair f and $\mathscr{C}$ induces natural cell decompositions $\mathbf{D}^n(f, \mathscr{C})$ of S^2, for $n \in \mathbb{N}_0$, in the following way:

By the Jordan curve theorem, the set $S^2 \setminus \mathscr{C}$ has two connected components. We call the closure of one of them the *white* 0-*tile* for $(f, \mathscr{C})$, denoted by X^0_w, and the closure of the other one the *black* 0-*tile* for $(f, \mathscr{C})$, denoted by X^0_b. The set of 0-*tiles* is $\mathbf{X}^0(f, \mathscr{C}) = \{X^0_b, X^0_w\}$. The set of 0-*vertices* is $\mathbf{V}^0(f, \mathscr{C}) = \operatorname{post} f$. We set $\overline{\mathbf{V}}^0(f, \mathscr{C}) = \{\{x\} \,|\, x \in \mathbf{V}^0(f, \mathscr{C})\}$. The set of 0-*edges* $\mathbf{E}^0(f, \mathscr{C})$ is the set of the closures of the connected components of $\mathscr{C} \setminus \operatorname{post} f$. Then we get a cell decomposition

$$\mathbf{D}^0(f, \mathscr{C}) = \mathbf{X}^0(f, \mathscr{C}) \cup \mathbf{E}^0(f, \mathscr{C}) \cup \overline{\mathbf{V}}^0(f, \mathscr{C})$$

of S^2 consisting of *cells of level* 0, or 0-*cells*.

We can recursively define unique cell decompositions $\mathbf{D}^n(f, \mathscr{C})$, $n \in \mathbb{N}$, consisting of n-*cells* such that f is cellular for $(\mathbf{D}^{n+1}(f, \mathscr{C}), \mathbf{D}^n(f, \mathscr{C}))$. We refer to [BM17, Lemma 5.15] for more details. We denote by $\mathbf{X}^n(f, \mathscr{C})$ the set of n-cells of dimension 2, called n-*tiles*; by $\mathbf{E}^n(f, \mathscr{C})$ the set of n-cells of dimension 1, called n-*edges*; by $\overline{\mathbf{V}}^n(f, \mathscr{C})$ the set of n-cells of dimension 0; and by $\mathbf{V}^n(f, \mathscr{C})$ the set $\{x \,|\, \{x\} \in \overline{\mathbf{V}}^n(f, \mathscr{C})\}$, called the set of n-*vertices*. The k-*skeleton*, for $k \in \{0, 1, 2\}$, of $\mathbf{D}^n(f, \mathscr{C})$ is the union of all n-cells of dimension k in this cell decomposition.

We record Proposition 5.17 of [BM17] here in order to summarize properties of the cell decompositions $\mathbf{D}^n(f, \mathscr{C})$ defined above.

Proposition 2.6 (M. Bonk & D. Meyer) *Let $k, n \in \mathbb{N}_0$, let $f : S^2 \to S^2$ be a Thurston map, $\mathscr{C} \subseteq S^2$ be a Jordan curve with* $\operatorname{post} f \subseteq \mathscr{C}$, *and* $m = \operatorname{card}(\operatorname{post} f)$.

(i) *The map f^k is cellular for $(\mathbf{D}^{n+k}(f, \mathscr{C}), \mathbf{D}^n(f, \mathscr{C}))$. In particular, if c is any $(n+k)$-cell, then $f^k(c)$ is an n-cell, and $f^k|_c$ is a homeomorphism of c onto $f^k(c)$.*
(ii) *Let c be an n-cell. Then $f^{-k}(c)$ is equal to the union of all $(n+k)$-cells c' with $f^k(c') = c$.*

(iii) *The* 1*-skeleton of* $\mathbf{D}^n(f,\mathscr{C})$ *is equal to* $f^{-n}(\mathscr{C})$. *The* 0*-skeleton of* $\mathbf{D}^n(f,\mathscr{C})$ *is the set* $\mathbf{V}^n(f,\mathscr{C}) = f^{-n}(\text{post } f)$, *and we have* $\mathbf{V}^n(f,\mathscr{C}) \subseteq \mathbf{V}^{n+k}(f,\mathscr{C})$.
(iv) $\text{card}(\mathbf{X}^n(f,\mathscr{C})) = 2(\deg f)^n$, $\text{card}(\mathbf{E}^n(f,\mathscr{C})) = m(\deg f)^n$, *and* $\text{card}(\mathbf{V}^n(f,\mathscr{C})) \leq m(\deg f)^n$.
(v) *The* n*-edges are precisely the closures of the connected components of* $f^{-n}(\mathscr{C}) \setminus f^{-n}(\text{post } f)$. *The* n*-tiles are precisely the closures of the connected components of* $S^2 \setminus f^{-n}(\mathscr{C})$.
(vi) *Every* n*-tile is an* m*-gon, i.e., the number of* n*-edges and the number of* n*-vertices contained in its boundary are equal to* m.

We also note that for each n-edge $e \in \mathbf{E}^n(f,\mathscr{C})$, $n \in \mathbb{N}_0$, there exist exactly two n-tiles $X, X' \in \mathbf{X}^n(f,\mathscr{C})$ such that $X \cap X' = e$.

For $n \in \mathbb{N}_0$, we define *the set of black* n*-tiles* as

$$\mathbf{X}^n_b(f,\mathscr{C}) = \{X \in \mathbf{X}^n(f,\mathscr{C}) \mid f^n(X) = X^0_b\},$$

and the *set of white* n*-tiles* as

$$\mathbf{X}^n_w(f,\mathscr{C}) = \{X \in \mathbf{X}^n(f,\mathscr{C}) \mid f^n(X) = X^0_w\}.$$

It follows immediately from Proposition 2.6 that

$$\text{card}\left(\mathbf{X}^n_b(f,\mathscr{C})\right) = \text{card}\left(\mathbf{X}^n_w(f,\mathscr{C})\right) = (\deg f)^n \tag{2.4}$$

for each $n \in \mathbb{N}_0$. Moreover, for $n \in \mathbb{N}$, we define the *set of black* n*-tiles contained in a white* $(n-1)$*-tile* as

$$\mathbf{X}^n_{bw}(f,\mathscr{C}) = \{X \in \mathbf{X}^n_b(f,\mathscr{C}) \mid \exists X' \in \mathbf{X}^{n-1}_w(f,\mathscr{C}),\ X \subseteq X'\},$$

the *set of black* n*-tiles contained in a black* $(n-1)$*-tile* as

$$\mathbf{X}^n_{bb}(f,\mathscr{C}) = \{X \in \mathbf{X}^n_b(f,\mathscr{C}) \mid \exists X' \in \mathbf{X}^{n-1}_b(f,\mathscr{C}),\ X \subseteq X'\},$$

the *set of white* n*-tiles contained in a black* $(n-1)$*-tile* as

$$\mathbf{X}^n_{wb}(f,\mathscr{C}) = \{X \in \mathbf{X}^n_w(f,\mathscr{C}) \mid \exists X' \in \mathbf{X}^{n-1}_b(f,\mathscr{C}),\ X \subseteq X'\},$$

and *the set of white* n*-tiles contained in a while* $(n-1)$*-tile* as

$$\mathbf{X}^n_{ww}(f,\mathscr{C}) = \{X \in \mathbf{X}^n_w(f,\mathscr{C}) \mid \exists X' \in \mathbf{X}^{n-1}_w(f,\mathscr{C}),\ X \subseteq X'\}.$$

In other words, for example, a black n-tile is an n-tile that is mapped by f^n to the black 0-tile, and a black n-tile contained in a white $(n-1)$-tile is an n-tile that is contained in some white $(n-1)$-tile as a set, and is mapped by f^n to the black 0-tile.

If we fix the cell decomposition $\mathbf{D}^n(f, \mathscr{C})$, $n \in \mathbb{N}_0$, we can define for each $v \in \mathbf{V}^n(f, \mathscr{C})$ the *n-flower of* v as

$$W^n(v) = \bigcup\{\text{inte}(c) \mid c \in \mathbf{D}^n(f, \mathscr{C}),\ v \in c\}. \tag{2.5}$$

Note that flowers are open (in the standard topology on S^2). Let $\overline{W}^n(v)$ be the closure of $W^n(v)$. We define the *set of all n-flowers* by

$$\mathbf{W}^n(f, \mathscr{C}) = \{W^n(v) \mid v \in \mathbf{V}^n(f, \mathscr{C})\}. \tag{2.6}$$

From now on, if the map f and the Jordan curve $\mathscr{C}$ are clear from the context, we will sometimes omit $(f, \mathscr{C})$ in the notation above.

Remark 2.7 For $n \in \mathbb{N}_0$ and $v \in \mathbf{V}^n$, we have

$$\overline{W}^n(v) = X_1 \cup X_2 \cup \cdots \cup X_m,$$

where $m = 2\deg_{f^n}(v)$, and $X_1, X_2, \ldots X_m$ are all the n-tiles that contain v as a vertex (see [BM17, Lemma 5.28]). Moreover, each flower is mapped under f to another flower in such a way that is similar to the map $z \mapsto z^k$ on the complex plane. More precisely, for $n \in \mathbb{N}_0$ and $v \in \mathbf{V}^{n+1}$, there exist orientation preserving homeomorphisms $\varphi\colon W^{n+1}(v) \to D$ and $\eta\colon W^n(f(v)) \to D$ such that D is the unit disk on $\mathbb{C}$, $\varphi(v) = 0$, $\eta(f(v)) = 0$, and

$$(\eta \circ f \circ \varphi^{-1})(z) = z^k$$

for all $z \in D$, where $k = \deg_f(v)$. Let $\overline{W}^{n+1}(v) = X_1 \cup X_2 \cup \cdots \cup X_m$ and $\overline{W}^n(f(v)) = X'_1 \cup X'_2 \cup \cdots \cup X'_{m'}$, where $X_1, X_2, \ldots X_m$ are all the $(n+1)$-tiles that contain v as a vertex, listed counterclockwise, and $X'_1, X'_2, \ldots X'_{m'}$ are all the n-tiles that contain $f(v)$ as a vertex, listed counterclockwise, and $f(X_1) = X'_1$. Then $m = m'k$, and $f(X_i) = X'_j$ if $i \equiv j \pmod k$, where $k = \deg_f(v)$. (See also Case 3 of the proof of Lemma 5.24 in [BM17] for more details.)

We denote, for each $x \in S^2$,

$$U^n(x) = \bigcup\{Y^n \in \mathbf{X}^n \mid \text{there exists } X^n \in \mathbf{X}^n \text{with } x \in X^n,\ X^n \cap Y^n \neq \emptyset\}, \tag{2.7}$$

and for each integer $m \leq -1$, set $U^m(x) = S^2$. We define the *n-partition* O_n of S^2 induced by $(f, \mathscr{C})$ as

$$O_n = \{\text{inte}(X^n) \mid X^n \in \mathbf{X}^n\} \cup \{\text{inte}(e^n) \mid e^n \in \mathbf{E}^n\} \cup \overline{\mathbf{V}}^n. \tag{2.8}$$

2.4 Notions of Expansion for Thurston Maps

We now define two notions of expansion introduced by M. Bonk and D. Meyer [BM17].

It is proved in [BM17, Corollary 7.2] that for each expanding Thurston map f (see Definition 2.10), we have $\operatorname{card}(\operatorname{post} f) \geq 3$.

Definition 2.8 (*Joining opposite sides*) Fix a Thurston map f with $\operatorname{card}(\operatorname{post} f) \geq 3$ and an f-invariant Jordan curve $\mathscr{C}$ containing $\operatorname{post} f$. A set $K \subseteq S^2$ *joins opposite sides* of $\mathscr{C}$ if K meets two disjoint 0-edges when $\operatorname{card}(\operatorname{post} f) \geq 4$, or K meets all three 0-edges when $\operatorname{card}(\operatorname{post} f) = 3$.

Definition 2.9 (*Combinatorial expansion*) Let f be a Thurston map. We say that f is *combinatorially expanding* if $\operatorname{card}(\operatorname{post} f) \geq 3$, and there exists an f-invariant Jordan curve $\mathscr{C} \subseteq S^2$ (i.e., $f(\mathscr{C}) \subseteq \mathscr{C}$) with $\operatorname{post} f \subseteq \mathscr{C}$, and there exists a number $n_0 \in \mathbb{N}$ such that none of the n_0-tiles in $\mathbf{X}^{n_0}(f, \mathscr{C})$ joins opposite sides of $\mathscr{C}$.

Definition 2.10 (*Expansion*) A Thurston map $f \colon S^2 \to S^2$ is called *expanding* if there exist a metric d on S^2 that induces the standard topology on S^2 and a Jordan curve $\mathscr{C} \subseteq S^2$ containing $\operatorname{post} f$ such that

$$\lim_{n \to +\infty} \max\{\operatorname{diam}_d(X) \mid X \in \mathbf{X}^n(f, \mathscr{C})\} = 0.$$

We call such a Thurston map an *expanding Thurston map*.

Remark 2.11 It is clear that if f is an expanding Thurston map, so is f^n for each $n \in \mathbb{N}$. We observe that being expanding is a topological property of a Thurston map and independent of the choice of the metric d that generates the standard topology on S^2. By Lemma 6.1 in [BM17], it is also independent of the choice of the Jordan curve $\mathscr{C}$ containing $\operatorname{post} f$. More precisely, if f is an expanding Thurston map, then

$$\lim_{n \to +\infty} \max\big\{\operatorname{diam}_{\widetilde{d}}(X) \,\big|\, X \in \mathbf{X}^n\big(f, \widetilde{\mathscr{C}}\big)\big\} = 0,$$

for each metric $\widetilde{d}$ that generates the standard topology on S^2 and each Jordan curve $\widetilde{\mathscr{C}} \subseteq S^2$ that contains $\operatorname{post} f$.

P. Haïssinsky and K. Pilgrim developed a notion of expansion in a more general context for finite branched coverings between topological spaces (see [HP09, Sect. 2.1 and Sect. 2.2]). This applies to Thurston maps and their notion of expansion is equivalent to our notion defined above in the context of Thurston maps (see [BM17, Proposition 6.3]). Such concepts of expansion are natural analogs, in the contexts of finite branched coverings and Thurston maps, to some of the more classical versions, such as expansive homeomorphisms and forward-expansive continuous maps between compact metric spaces (see for example, [KH95, Definition 3.2.11]), and distance-expanding maps between compact metric spaces (see for example, [PU10,

Chap. 4]). Our notion of expansion is not equivalent to any such classical notion in the context of Thurston maps. In fact, as mentioned in the introduction, there are subtle connections between our notion of expansion and some classical notions of weak expansion. Chapter 6 will be devoted to this topic. See Theorem 6.1 for the precise statement.

Lemma 2.12 *Let $f\colon S^2 \to S^2$ be an expanding Thurston map. Then for each $p \in S^2$, the set $\bigcup_{n=1}^{+\infty} f^{-n}(p)$ is dense in S^2, and*

$$\lim_{n\to+\infty} \operatorname{card}(f^{-n}(p)) = +\infty. \tag{2.9}$$

Proof Let $\mathscr{C} \subseteq S^2$ be a Jordan curve containing post f. Let d be any metric on S^2 that generates the standard topology on S^2.

Without loss of generality, we assume that $p \in X^0_w$ where $X^0_w \in \mathbf{X}^0_w(f, \mathscr{C})$ is the white 0-tile in the cell decompositions induced by $(f, \mathscr{C})$. The proof for the case when $p \in X^0_b$ where $X^0_b \in \mathbf{X}^0_b(f, \mathscr{C})$ is the black 0-tile is similar.

By Proposition 2.6(ii), for each $n \in \mathbb{N}$ and each white n-tile $X^n_w \in \mathbf{X}^n_w(f, \mathscr{C})$, there is a point $q \in X^n_w$ with $f^n(q) = p$. Since f is an expanding Thurston map,

$$\lim_{n\to+\infty} \max\{\operatorname{diam}_d(X) \mid X \in \mathbf{X}^n(f, \mathscr{C})\} = 0. \tag{2.10}$$

Then the density of the set $\bigcup_{n=1}^{+\infty} f^{-n}(p)$ follows from the observation that for each $n \in \mathbb{N}$, each black n-tile $X^n_b \in \mathbf{X}^n_b(f, \mathscr{C})$ intersects nontrivially with some white n-tile $X^n_w \in \mathbf{X}^n_w(f, \mathscr{C})$.

By the above observation, the triangular inequality, and the fact that $\operatorname{diam}_d(S^2) > 0$ and S^2 is connected in the standard topology, the equation (2.9) follows from (2.10). □

2.5 Visual Metric

For an expanding Thurston map f, we can fix a particular metric d on S^2 called *visual metric for f*. For the existence and properties of such metrics, see [BM17, Chap. 8]. For a visual metric d for f, there exists a unique constant $\Lambda > 1$ called the *expansion factor* of d (see [BM17, Chap. 8] for more details). One major advantage of a visual metric d is that in (S^2, d) we have good quantitative control over the sizes of the cells in the cell decompositions discussed above. We summarize several results of this type ([BM17, Proposition 8.4, Lemmas 8.10, 8.11]) in the lemma below.

Lemma 2.13 (M. Bonk and D. Meyer) *Let $f\colon S^2 \to S^2$ be an expanding Thurston map, and $\mathscr{C} \subseteq S^2$ be a Jordan curve containing* post f. *Let d be a visual metric on*

S^2 for f with expansion factor $\Lambda > 1$. Then there exist constants $C \geq 1$, $C' \geq 1$, $K \geq 1$, and $n_0 \in \mathbb{N}_0$ with the following properties:

(i) *$d(\sigma, \tau) \geq C^{-1}\Lambda^{-n}$ whenever σ and τ are disjoint n-cells for $n \in \mathbb{N}_0$.*
(ii) *$C^{-1}\Lambda^{-n} \leq \operatorname{diam}_d(\tau) \leq C\Lambda^{-n}$ for all n-edges and all n-tiles τ for $n \in \mathbb{N}_0$.*
(iii) *$B_d(x, K^{-1}\Lambda^{-n}) \subseteq U^n(x) \subseteq B_d(x, K\Lambda^{-n})$ for $x \in S^2$ and $n \in \mathbb{N}_0$.*
(iv) *$U^{n+n_0}(x) \subseteq B_d(x, r) \subseteq U^{n-n_0}(x)$ where $n = \lceil -\log r / \log \Lambda \rceil$ for $r > 0$ and $x \in S^2$.*
(v) *For every n-tile $X^n \in \mathbf{X}^n(f, \mathscr{C})$, $n \in \mathbb{N}_0$, there exists a point $p \in X^n$ such that $B_d(p, C^{-1}\Lambda^{-n}) \subseteq X^n \subseteq B_d(p, C\Lambda^{-n})$.*

Conversely, if $\widetilde{d}$ is a metric on S^2 satisfying conditions (i) *and* (ii) *for some constant $C \geq 1$, then $\widetilde{d}$ is a visual metric with expansion factor $\Lambda > 1$.*

Recall $U^n(x)$ is defined in (2.7).

In addition, we will need the fact that a visual metric d induces the standard topology on S^2 ([BM17, Proposition 8.3]) and the fact that the metric space (S^2, d) is linearly locally connected ([BM17, Proposition 18.5]). A metric space (X, d) is *linearly locally connected* if there exists a constant $L \geq 1$ such that the following conditions are satisfied:

1. For all $z \in X$, $r > 0$, and $x, y \in B_d(z, r)$ with $x \neq y$, there exists a continuum $E \subseteq X$ with $x, y \subseteq E$ and $E \subseteq B_d(z, rL)$.
2. For all $z \in X$, $r > 0$, and $x, y \in X \setminus B_d(z, r)$ with $x \neq y$, there exists a continuum $E \subseteq X$ with $x, y \subseteq E$ and $E \subseteq X \setminus B_d(z, r/L)$.

We call such a constant $L \geq 1$ a *linear local connectivity constant of d*.

Remark 2.14 If $f : \widehat{\mathbb{C}} \to \widehat{\mathbb{C}}$ is a rational expanding Thurston map, then a visual metric is quasisymmetrically equivalent to the chordal metric on the Riemann sphere $\widehat{\mathbb{C}}$ (see [BM17, Lemma 18.10]). Here the chordal metric σ on $\widehat{\mathbb{C}}$ is given by $\sigma(z, w) = \frac{2|z-w|}{\sqrt{1+|z|^2}\sqrt{1+|w|^2}}$ for $z, w \in \mathbb{C}$, and $\sigma(\infty, z) = \sigma(z, \infty) = \frac{2}{\sqrt{1+|z|^2}}$ for $z \in \mathbb{C}$. We also note that a quasisymmetric embedding of a bounded connected metric space is Hölder continuous (see [He01, Sect. 11.1 and Corollary 11.5]). Consequently, the classes of Hölder continuous functions on $\widehat{\mathbb{C}}$ equipped with the chordal metric and on $S^2 = \widehat{\mathbb{C}}$ equipped with any visual metric for f are the same (upto a change of the Hölder exponent).

An expanding Thurston map is Lipschitz with respect to a visual metric.

Lemma 2.15 *Let $f : S^2 \to S^2$ be an expanding Thurston map, and d be a visual metric on S^2 for f with expansion factor $\Lambda > 1$. Then f is Lipschitz with respect to d.*

Proof Fix a Jordan curve $\mathscr{C} \subseteq S^2$ containing post f. Let $x, y \in S^2$ and we assume that

$$0 < d(x, y) < K^{-1}\Lambda^{-2}, \tag{2.11}$$

where $K \geq 1$ is a constant from Lemma 2.13 depending only on f, $\mathscr{C}$, and d.

Set $m = \max\{k \in \mathbb{N}_0 \mid y \in U^k(x)\}$, where $U^k(x)$ is defined in (2.7). By Lemma 2.13(iii), the number m is finite. Then $y \notin U^{m+1}(x)$. Thus by Lemma 2.13(iii),

$$\frac{1}{K}\Lambda^{-m-1} \leq d(x, y) \leq K\Lambda^{-m}.$$

By (2.11) we get $m \geq 1$. Since $f(y) \in f(U^m(x)) \subseteq U^{m-1}(f(x))$ by Proposition 2.6, we get from Lemma 2.13(iii) that

$$d(f(x), f(y)) \leq K\Lambda^{-m+1}.$$

Therefore,

$$\frac{d(f(x), f(y))}{d(x, y)} \leq \frac{K\Lambda^{-m+1}}{\frac{1}{K}\Lambda^{-m-1}} = K^2\Lambda^2,$$

and f is Lipschitz with respect to d. □

2.6 Invariant Curves

A Jordan curve $\mathscr{C} \subseteq S^2$ is f-invariant if $f(\mathscr{C}) \subseteq \mathscr{C}$. We are interested in f-invariant Jordan curves that contain post f, since for such a curve $\mathscr{C}$, the partition $(\mathbf{D}^1(f, \mathscr{C}), \mathbf{D}^0(f, \mathscr{C}))$ is then a cellular Markov partition for f. According to Example 15.11 and Lemma 15.12 in [BM17], f-invariant Jordan curves containing post f need not exist. However, M. Bonk and D. Meyer [BM17, Theorem 15.1] proved that there exists an f^n-invariant Jordan curve $\mathscr{C}$ containing post f for each sufficiently large n depending on f.

Theorem 2.16 (M. Bonk & D. Meyer) *Let $f: S^2 \to S^2$ be an expanding Thurston map. Then for each $n \in \mathbb{N}$ sufficiently large, there exists a Jordan curve $\mathscr{C} \subseteq S^2$ containing* post f *such that $f^n(\mathscr{C}) \subseteq \mathscr{C}$.*

We will need a slightly stronger version in Chaps. 4 and 6. Its proof is almost the same as that of [BM17, Theorem 15.1]. For the convenience of the reader, we include the proof here.

Lemma 2.17 *Let $f: S^2 \to S^2$ be an expanding Thurston map, and $\widetilde{\mathscr{C}} \subseteq S^2$ be a Jordan curve with* post $f \subseteq \widetilde{\mathscr{C}}$. *Then there exists an integer $N(f, \widetilde{\mathscr{C}}) \in \mathbb{N}$ such that for each $n \geq N(f, \widetilde{\mathscr{C}})$ there exists an f^n-invariant Jordan curve $\mathscr{C}$ isotopic to $\widetilde{\mathscr{C}}$ rel.* post f *such that no n-tile in $\mathbf{D}^n(f, \mathscr{C})$ joins opposite sides of $\mathscr{C}$.*

Proof By [BM17, Lemma 15.15], there exists an integer $N(f, \widetilde{\mathscr{C}}) \in \mathbb{N}$ such that for each $n \geq N(f, \widetilde{\mathscr{C}})$, there exists a Jordan curve $\mathscr{C}' \subseteq f^{-n}(\widetilde{\mathscr{C}})$ that is isotopic to $\widetilde{\mathscr{C}}$ rel. post f, and no n-tile for $(f, \widetilde{\mathscr{C}})$ joins opposite sides of $\mathscr{C}'$. Let $H: S^2 \times [0, 1] \to S^2$

be this isotopy rel. post f. We set $H_t(x) = H(x,t)$ for $x \in S^2$, $t \in [0,1]$. We have $H_0 = \mathrm{id}_{S^2}$ and $\mathscr{C}' = H_1(\widetilde{\mathscr{C}}) \subseteq f^{-n}(\widetilde{\mathscr{C}})$.

If $F = f^n$, then post F = post f and F is also an expanding Thurston map ([BM17, Lemma 6.4]). Note that F is cellular for $(\mathbf{D}^n(f,\widetilde{\mathscr{C}}), \mathbf{D}^0(f,\widetilde{\mathscr{C}}))$. So $\mathbf{D}^1(F,\widetilde{\mathscr{C}}) = \mathbf{D}^n(f,\widetilde{\mathscr{C}})$ (see [BM17, Lemma 5.12]). Thus no 1-cell for $(H_1 \circ F, \mathscr{C}')$ joins opposite sides of $\mathscr{C}'$, and thus $H_1 \circ F$ is combinatorially expanding for $\mathscr{C}'$. Note that $\mathscr{C}'$ contains $\mathrm{post}(H_1 \circ F) = \mathrm{post}\, F = \mathrm{post}\, f$. By Theorem 14.2 in [BM17], there exists a homeomorphism $\phi\colon S^2 \to S^2$ that is isotopic to the identity rel. post $(H_1 \circ F)$ such that $\phi(\mathscr{C}') = \mathscr{C}'$ and $G = \phi \circ H_1 \circ F$ is an expanding Thurston map. Since $\phi \circ H_1$ is isotopic to the identity on S^2 rel. post F, the pair F and G are Thurston equivalent. By Theorem 11.1 in [BM17], there exists a homeomorphism $h\colon S^2 \to S^2$ that is isotopic to the identity on S^2 rel. $F^{-1}(\mathrm{post}\, F)$ with $F \circ h = h \circ G$. Set $\mathscr{C} = h(\mathscr{C}')$. Then $\mathscr{C}$ is a Jordan curve in S^2 that is isotopic to $\mathscr{C}'$ rel. $F^{-1}(\mathrm{post}\, F)$ and thus isotopic to $\widetilde{\mathscr{C}}$ rel. post F. Since $F(\mathscr{C}) = F(h(\mathscr{C}')) = h(G(\mathscr{C}')) = h(\phi(H_1(F(\mathscr{C}')))) \subseteq h(\phi(\mathscr{C}')) = h(\mathscr{C}') = \mathscr{C}$, we get that $\mathscr{C}$ is F-invariant.

Moreover, since no 1-cell for $(H_1 \circ F, \mathscr{C}')$ joins opposite sides of $\mathscr{C}'$, $H_1 \circ F(\mathscr{C}') \subseteq H_1(\widetilde{\mathscr{C}}) = \mathscr{C}'$, $\phi\colon S^2 \to S^2$ is a homeomorphism isotopic to the identity rel. $\mathrm{post}(H_1 \circ F)$ with $\phi(\mathscr{C}') = \mathscr{C}'$, $G = \phi \circ H_1 \circ F$, we can conclude that $G(\mathscr{C}') \subseteq \mathscr{C}'$ and no 1-cell for $(G, \mathscr{C}')$ joins opposite sides of $\mathscr{C}'$. Since $h\colon S^2 \to S^2$ is a homeomorphism, $\mathscr{C} = h(\mathscr{C}')$, and $F \circ h = h \circ G$, we can finally conclude that no 1-cell for $(F, \mathscr{C})$ joins opposite sides of $\mathscr{C}$. Therefore no n-cell for $(f, \mathscr{C})$ joins opposite sides of $\mathscr{C}$. □

Compared with [BM17, Theorem 15.1], the above lemma carries additional combinatorial information of $\mathscr{C}$, i.e., no n-tile joins opposite sides of $\mathscr{C}$. In fact, we will only need the following corollary of Lemma 2.17 in Chaps. 4 and 6.

Corollary 2.18 *Let $f\colon S^2 \to S^2$ be an expanding Thurston map. Then there exists a constant $N(f) > 0$ such that for each $n \geq N(f)$, there exists an f^n-invariant Jordan curve $\mathscr{C}$ containing* post f *such that no n-tile in $\mathbf{D}^n(f,\mathscr{C})$ joins opposite sides of $\mathscr{C}$.*

Proof We can choose an arbitrary Jordan curve $\widetilde{\mathscr{C}} \subseteq S^2$ containing post f and set $N(f) = N(f,\widetilde{\mathscr{C}})$, and $\mathscr{C}$ an f^n-invariant Jordan curve containing post f as in Lemma 2.17. □

We now establish a generalization of [BM17, Lemma 15.25]. It is an essential ingredient for the distortion lemmas (Lemma 5.3 and Lemma 5.4) that we will repeatedly use in Chaps. 5 and 7.

Lemma 2.19 *Let $f\colon S^2 \to S^2$ be an expanding Thurston map, and $\mathscr{C} \subseteq S^2$ be a Jordan curve that satisfies* post $f \subseteq \mathscr{C}$ *and $f^{n_\mathscr{C}}(\mathscr{C}) \subseteq \mathscr{C}$ for some $n_\mathscr{C} \in \mathbb{N}$. Let d be a visual metric on S^2 for f with expansion factor $\Lambda > 1$. Then there exists a constant $C_0 > 1$, depending only on f, d, $\mathscr{C}$, and $n_\mathscr{C}$, with the following property:*

If $k, n \in \mathbb{N}_0$, $X^{n+k} \in \mathbf{X}^{n+k}(f, \mathcal{C})$, *and* $x, y \in X^{n+k}$, *then*

$$\frac{1}{C_0} d(x, y) \le \frac{d(f^n(x), f^n(y))}{\Lambda^n} \le C_0 d(x, y). \tag{2.12}$$

Proof In this proof, we set a constant $K = 2\max\{1, l_f\}$, where l_f is the Lipschitz constant of f with respect to d. Let $N = n_{\mathcal{C}}$.

By Remark 2.11, the map f^N is an expanding Thurston map. It is easy to see from Lemma 2.13 that the metric d is a visual metric for the expanding Thurston map f^N with expansion factor Λ^N. So by Lemma 15.25 in [BM17], there exists a constant $D \ge 1$ depending only on f^N, $\mathcal{C}$, and d such that for each $k, l \in \mathbb{N}_0$, each $X \in \mathbf{X}^{(l+k)N}(f, \mathcal{C})$, and each pair of points $x, y \in X$, we have

$$\frac{1}{D} d(x, y) \le \frac{d(f^{lN}(x), f^{lN}(y))}{\Lambda^{lN}} \le D d(x, y). \tag{2.13}$$

Fix $m, l \in \mathbb{N}_0$, $s, t \in \{0, 1, \dots, N-1\}$, $X \in \mathbf{X}^{(mN+s)+(lN+t)}(f, \mathcal{C})$, and $x, y \in X$.

We prove the second inequality in (2.12) with $n = mN + s$ and $k = lN + t$ by considering the following cases depending on whether $l = 0$ or $l \ge 1$.

If $l = 0$, then by Lemma 2.15 and the fact that $K > l_f$,

$$d\left(f^{lN+t}(x), f^{lN+t}(y)\right) \le K^t d(x, y) \le K^{2N} d(x, y) \Lambda^{lN+t}.$$

If $l \ge 1$, then by Lemma 2.15, (2.13), and the fact that $K > l_f$,

$$\begin{aligned}
& d\left(f^{lN+t}(x), f^{lN+t}(y)\right) \\
=& d\left(f^{(l-1)N+(N-s)}\left(f^{t+s}(x)\right), f^{(l-1)N+(N-s)}\left(f^{t+s}(y)\right)\right) \\
\le& K^{N-s} d\left(f^{(l-1)N}\left(f^{t+s}(x)\right), f^{(l-1)N}\left(f^{t+s}(y)\right)\right) \\
\le& K^{N-s} D d\left(f^{t+s}(x), f^{t+s}(y)\right) \Lambda^{(l-1)N} \\
\le& K^{N-s} D\left(K^{t+s} d(x, y)\right) \Lambda^{lN+t} \\
\le& K^{2N} D d(x, y) \Lambda^{lN+t}.
\end{aligned}$$

We consider the first inequality in (2.12) with $n = mN + s$ and $k = lN + t$ now. By Proposition 2.6(i), we can choose $Y \in \mathbf{X}^{(m+l+2)N}(f, \mathcal{C})$ and two points $x', y' \in Y$ such that $f^{2N-s-t}(Y) = X$, $f^{2N-s-t}(x') = x$, and $f^{2N-s-t}(y') = y$. Note that $2N - s - t \ge 2$. Then by Lemma 2.15, (2.13), and the fact that $K > l_f$,

$$\begin{aligned}
& d\left(f^{lN+t}(x), f^{lN+t}(y)\right) \\
=& d\left(f^{lN+t}\left(f^{2N-s-t}(x')\right), f^{lN+t}\left(f^{2N-s-t}(y')\right)\right) \\
=& d\left(f^{lN+2N-s}(x'), f^{lN+2N-s}(y')\right) \\
\ge& K^{-s} d\left(f^{lN+2N}(x'), f^{lN+2N}(y')\right)
\end{aligned}$$

$$
\begin{aligned}
&\geq K^{-s} D^{-1} d(x', y') \Lambda^{lN+2N} \\
&\geq K^{-s} D^{-1} K^{-(2N-s-t)} d(x, y) \Lambda^{lN+t} \\
&\geq K^{-2N} D^{-1} d(x, y) \Lambda^{lN+t}.
\end{aligned}
$$

Therefore,

$$
\frac{1}{C_0} d(x, y) \leq \frac{d(f^{lN+t}(x), f^{lN+t}(y))}{\Lambda^{lN+t}} \leq C_0 d(x, y),
$$

where $C_0 = K^{2N} D$ is a constant depending only on $f, d, \mathscr{C}$, and $N = n_{\mathscr{C}}$. □

Chapter 3
Ergodic Theory

In this chapter, we review some key concepts of ergodic theory. The usual notions of covers and partitions are introduced in Sect. 3.1. Then in Sect. 3.2, measure-theoretic entropy and topological entropy, as well as measure-theoretic pressure and topological pressure are introduced before measures of maximal entropy and equilibrium states are defined. In Sect. 3.3, we formulate the Ruelle operator $\mathscr{L}_{\psi}$ for an expanding Thurston map and a complex-valued continuous function ψ on S^2, which will is the key tool and object of investigation in thermodynamical formalism in Chap. 5. We argue that it is well-defined on the space of real-valued continuous functions on S^2. We then discuss some of the properties of the Ruelle operator. In Sect. 3.4, we review the notion of topological conditional entropy $h(g|\lambda)$ of a continuous map $g : X \to X$ (on a compact metric space X) given an open cover λ of X, and the notion of topological tail entropy $h^*(g)$ of g. The latter was first introduced by M. Misiurewicz under the name "topological conditional entropy" [Mis73, Mis76]. We adopt the terminology and formulations by T. Downarowicz in [Dow11]. By using these notions, we then define h-expansiveness and asymptotic h-expansiveness, which will be used in Chaps. 6 and 7.

3.1 Covers and Partitions

Let (X, d) be a compact metric space and $g: X \to X$ a continuous map.

A *cover* of X is a collection $\xi = \{A_j \mid j \in J\}$ of subsets of X with the property that $\bigcup \xi = X$, where J is an index set. The cover ξ is an *open cover* if A_j is an open set for each $j \in J$. The cover ξ is *finite* if the index set J is a finite set.

A *measurable partition* ξ of X is a cover $\xi = \{A_j \mid j \in J\}$ of X consisting of countably many mutually disjoint Borel sets A_j, $j \in J$, where J is a countable index set. For $x \in X$, we denote by $\xi(x)$ the unique element of ξ that contains x.

Z. Li, *Ergodic Theory of Expanding Thurston Maps*, Atlantis Studies in Dynamical Systems 4, DOI 10.2991/978-94-6239-174-1_3

Let $\xi = \{A_j \mid j \in J\}$ and $\eta = \{B_k \mid k \in K\}$ be two covers of X, where J and K are the corresponding index sets. We say ξ is a *refinement* of η if for each $A_j \in \xi$, there exists $B_k \in \eta$ such that $A_j \subseteq B_k$. The *common refinement* $\xi \vee \eta$ of ξ and η defined as

$$\xi \vee \eta = \{A_j \cap B_k \mid j \in J,\ k \in K\}$$

is also a cover. Note that if ξ and η are both open covers (resp., measurable partitions), then $\xi \vee \eta$ is also an open cover (resp., a measurable partition). Define $g^{-1}(\xi) = \{g^{-1}(A_j) \mid j \in J\}$, and denote for $n \in \mathbb{N}$,

$$\xi_g^n = \bigvee_{j=0}^{n-1} g^{-j}(\xi) = \xi \vee g^{-1}(\xi) \vee \cdots \vee g^{-(n-1)}(\xi),$$

and let ξ_g^∞ be the smallest σ-algebra containing $\bigcup\limits_{n=1}^{+\infty} \xi_g^n$.

We adopt the following definition from [Dow11, Remark 6.1.7].

Definition 3.1 (*Refining sequences of open covers*) A sequence of open covers $\{\xi_i\}_{i\in\mathbb{N}_0}$ of a compact metric space X is a *refining sequence of open covers* of X if the following conditions are satisfied

(i) ξ_{i+1} is a refinement of ξ_i for each $i \in \mathbb{N}_0$.
(ii) For each open cover η of X, there exists $j \in \mathbb{N}$ such that ξ_i is a refinement of η for each $i \geq j$.

By the Lebesgue Number Lemma [Mu00, Lemma 27.5], it is clear that for a compact metric space, refining sequences of open covers always exist.

3.2 Entropy and Pressure

Let (X, d) be a compact metric space and $g\colon X \to X$ a continuous map. For $n \in \mathbb{N}$ and $x, y \in X$,

$$d_g^n(x, y) = \max\left\{d\big(g^k(x), g^k(y)\big) \big| k \in \{0, 1, \ldots, n-1\}\right\}$$

defines a new metric on X. A set $F \subseteq X$ is (n, ε)-separated, for some $n \in \mathbb{N}$ and $\varepsilon > 0$, if for each pair of distinct points $x, y \in F$, we have $d_g^n(x, y) \geq \varepsilon$. For $\varepsilon > 0$ and $n \in \mathbb{N}$, let $F_n(\varepsilon)$ be a maximal (in the sense of inclusion) (n, ε)-separated set in X.

For each $\psi \in C(X)$, the following limits exist and are equal, and we denote the limits by $P(g, \psi)$ (see for example, [PU10, Theorem 3.3.2]):

$$P(g,\psi) = \lim_{\varepsilon\to 0}\limsup_{n\to+\infty}\frac{1}{n}\log\sum_{x\in F_n(\varepsilon)}\exp(S_n\psi(x))$$
$$= \lim_{\varepsilon\to 0}\liminf_{n\to+\infty}\frac{1}{n}\log\sum_{x\in F_n(\varepsilon)}\exp(S_n\psi(x)), \tag{3.1}$$

where $S_n\psi(x) = \sum_{j=0}^{n-1}\psi(g^j(x))$ is defined in (0.3). We call $P(g,\psi)$ the *topological pressure* of g with respect to the *potential* ψ. The quantity $h_{\text{top}}(g) = P(g,0)$ is called the *topological entropy* of g. Note that $P(g,\psi)$ is independent of d as long as the topology on X defined by d remains the same (see [PU10, Sect. 3.2]).

We now review measure-theoretic counterparts of the concepts above.

The *information function* I maps a measurable partition ξ of X to a μ-a.e. defined real-valued function on X in the following way:

$$I(\xi)(x) = -\log\mu(\xi(x)), \qquad \text{for } x\in X. \tag{3.2}$$

Here $\xi(x)$ denotes the unique element of ξ that contains x.

Let ξ be a measurable partition of X. The *entropy* of ξ is

$$H_\mu(\xi) = -\sum_{j\in J}\mu(A_j)\log\left(\mu(A_j)\right),$$

where $0\log 0$ is defined to be 0. One can show (see [Wal82, Chap. 4]) that if $H_\mu(\xi) < +\infty$, then the following limit exists:

$$h_\mu(g,\xi) = \lim_{n\to+\infty}\frac{1}{n}H_\mu(\xi_g^n)\in[0,+\infty).$$

The *measure-theoretic entropy* of g for μ is given by

$$h_\mu(g) = \sup\{h_\mu(g,\xi)\,|\,\xi \text{ is a measurable partition of } X \text{ with } H_\mu(\xi) < +\infty\}. \tag{3.3}$$

For each $\psi\in C(X)$, the *measure-theoretic pressure* $P_\mu(g,\psi)$ of g for the measure μ and the potential ψ is

$$P_\mu(g,\psi) = h_\mu(g) + \int\psi\,\mathrm{d}\mu. \tag{3.4}$$

By the Variational Principle (see for example, [PU10, Theorem 3.4.1]), we have that for each $\psi\in C(X)$,

$$P(g,\psi) = \sup\{P_\mu(g,\psi)\,|\,\mu\in\mathcal{M}(X,g)\}. \tag{3.5}$$

In particular, when ψ is the constant function 0,

$$h_{\text{top}}(g) = \sup\{h_\mu(g) \mid \mu \in \mathcal{M}(X, g)\}. \tag{3.6}$$

A measure μ that attains the supremum in (3.5) is called an *equilibrium state* for the transformation g and the potential ψ. A measure μ that attains the supremum in (3.6) is called a *measure of maximal entropy* of g.

3.3 The Ruelle Operator for Expanding Thurston Maps

Let $f\colon S^2 \to S^2$ be an expanding Thurston map and $\psi \in C(S^2, \mathbb{C})$ a continuous function. We define the *Ruelle operator* $\mathcal{L}_\psi$ on $C(S^2, \mathbb{C})$ as the following

$$\mathcal{L}_\psi(u)(x) = \sum_{y \in f^{-1}(x)} \deg_f(y) u(y) \exp(\psi(y)), \tag{3.7}$$

for each $u \in C(S^2, \mathbb{C})$. To show that $\mathcal{L}_\psi$ is well-defined, we need to prove that $\mathcal{L}_\psi(u)(x)$ is continuous in $x \in S^2$ for each $u \in C(S^2, \mathbb{C})$. Indeed, by fixing an arbitrary Jordan curve $\mathcal{C} \subseteq S^2$ containing $\operatorname{post} f$, we know that for each x in the white 0-tile X^0_w,

$$\mathcal{L}_\psi(u)(x) = \sum_{X \in \mathbf{X}^1_w} u(y_X) \exp(\psi(y_X)),$$

where y_X is the unique point contained in the white 1-tile X with the property that $f(y_X) = x$ (Proposition 2.6(i)). If we move x around continuously within X^0_w, then y_X moves around continuously within X for each $X \in \mathbf{X}^1_w$. Thus $\mathcal{L}_\psi(u)(x)$ restricted to X^0_w is continuous in x. Similarly, $\mathcal{L}_\psi(u)(x)$ restricted to the black 0-tile X^0_b is also continuous in x. Hence $\mathcal{L}_\psi(u)(x)$ is continuous in $x \in S^2$.

Note that by a similar argument as above, we see that the Ruelle operator $\mathcal{L}_\psi : C(S^2, \mathbb{C}) \to C(S^2, \mathbb{C})$ has a natural extension to the space of complex-valued bounded Borel functions $B(S^2, \mathbb{C})$ (equipped with the uniform norm) given by (3.7) for each $u \in B(S^2, \mathbb{C})$. We also observe that if $\psi \in C(S^2)$ is real-valued, then $\mathcal{L}_\psi(C(S^2)) \subseteq C(S^2)$ and $\mathcal{L}_\psi(B(S^2)) \subseteq B(S^2)$. Moreover, we note that by induction and (2.2) we have

$$\mathcal{L}^n_\psi(u)(x) = \sum_{y \in f^{-n}(x)} \deg_{f^n}(y) u(y) \exp(S_n\psi(y)), \tag{3.8}$$

and

$$\mathcal{L}_\psi(u(v \circ f))(x) = \sum_{y \in f^{-1}(x)} \deg_f(y) u(y)(v \circ f)(y) \exp(\psi(y)) = v(x)\mathcal{L}_\psi(u)(x), \tag{3.9}$$

for $u, v \in B(S^2, \mathbb{C})$, $x \in S^2$, and $n \in \mathbb{N}$.

We assume now that $\psi \in C(S^2)$ is real-valued. Then it is clear that $\mathcal{L}_\psi$ is a positive, continuous operator on $C(S^2)$ (resp. $B(S^2)$) with the operator norm $\sup\{\mathcal{L}_\psi(\mathbb{1})(x) \mid x \in S^2\}$. Recall that the adjoint operator $\mathcal{L}_\psi^* \colon C^*(S^2) \to C^*(S^2)$ of $\mathcal{L}_\psi$ acts on the dual space $C^*(S^2)$ of the Banach space $C(S^2)$. We identify $C^*(S^2)$ with the space $\mathcal{M}(S^2)$ of finite signed Borel measures on S^2 by the Riesz representation theorem. From now on, we write $\langle \mu, u \rangle = \int u \, d\mu$ whenever $u \in B(S^2)$ and $\mu \in \mathcal{M}(S^2)$.

Lemma 3.2 *Let $f \colon S^2 \to S^2$ be an expanding Thurston map, $\psi \in C(S^2)$, and $\mu \in C^*(S^2)$. Then*

(i) $\langle \mathcal{L}_\psi^*(\mu), u \rangle = \langle \mu, \mathcal{L}_\psi(u) \rangle$ *for* $u \in B(S^2)$.
(ii) For each Borel set $A \subseteq S^2$ on which f is injective, we have that $f(A)$ is a Borel set, and

$$\mathcal{L}_\psi^*(\mu)(A) = \int_{f(A)} (\deg_f(\cdot) \exp(\psi)) \circ (f|_A)^{-1} \, d\mu. \tag{3.10}$$

Recall that a collection $\mathfrak{P}$ of subsets of a set Ω is a *π-system* if it is closed under intersection, i.e., if $A, B \in \mathfrak{P}$ then $A \cap B \in \mathfrak{P}$. A collection $\mathfrak{L}$ of subsets of Ω is a *λ-system* if the following are satisfied: (1) $\Omega \in \mathfrak{L}$. (2) If $B, C \in \mathfrak{L}$ and $B \subseteq C$, then $C \setminus B \in \mathfrak{L}$. (3) If $A_n \in \mathfrak{L}$, $n \in \mathbb{N}$, with $A_n \subseteq A_{n+1}$, then $\bigcup\limits_{n \in \mathbb{N}} A_n \in \mathfrak{L}$.

Proof For (i), it suffices to show that for each Borel set $A \subseteq S^2$,

$$\langle \mathcal{L}_\psi^*(\mu), \mathbb{1}_A \rangle = \langle \mu, \mathcal{L}_\psi(\mathbb{1}_A) \rangle. \tag{3.11}$$

Let $\mathfrak{L}$ be the collection of Borel sets $A \subseteq S^2$ for which (3.11) holds. Denote the collection of open subsets of S^2 by $\mathfrak{G}$. Then $\mathfrak{G}$ is a π-system.

We first observe from (3.7) that if $\{u_n\}_{n \in \mathbb{N}}$ is a non-decreasing sequence of real-valued functions on S^2, then so is $\{\mathcal{L}_\psi(u_n)\}_{n \in \mathbb{N}}$.

By the definition of $\mathcal{L}_\psi^*$, we have

$$\langle \mathcal{L}_\psi^*(\mu), u \rangle = \langle \mu, \mathcal{L}_\psi(u) \rangle \tag{3.12}$$

for $u \in C(S^2)$. Fix an open set $U \subseteq S^2$, then there exists a non-decreasing sequence $\{g_n\}_{n \in \mathbb{N}}$ of real-valued continuous functions on S^2 supported in U such that g_n converges to $\mathbb{1}_U$ pointwise as $n \longrightarrow +\infty$. Then $\{\mathcal{L}_\psi(g_n)\}_{n \in \mathbb{N}}$ is also a non-decreasing sequence of continuous functions, whose pointwise limit is $\mathcal{L}_\psi(\mathbb{1}_U)$. By the Lebesgue Monotone Convergence Theorem and (3.12), we can conclude that (3.11) holds for $A = U$. Thus $\mathfrak{G} \subseteq \mathfrak{L}$.

We now prove that $\mathfrak{L}$ is a λ-system. Indeed, since (3.12) holds for $u = \mathbb{1}_{S^2}$, we get $S^2 \in \mathfrak{L}$. Given $B, C \in \mathfrak{L}$ with $B \subseteq C$, then $\mathbb{1}_C - \mathbb{1}_B = \mathbb{1}_{C \setminus B}$ and $\mathcal{L}_\psi(\mathbb{1}_C) - \mathcal{L}_\psi(\mathbb{1}_B) = \mathcal{L}_\psi(\mathbb{1}_C - \mathbb{1}_B) = \mathcal{L}_\psi(\mathbb{1}_{C \setminus B})$ by (3.7). Thus $C \setminus B \in \mathfrak{L}$. Finally, given $A_n \in \mathfrak{L}$, $n \in \mathbb{N}$, with $A_n \subseteq A_{n+1}$, and let $A = \bigcup\limits_{n \in \mathbb{N}} A_n$. Then $\{\mathbb{1}_{A_n}\}_{n \in \mathbb{N}}$ and $\{\mathcal{L}_\psi(\mathbb{1}_{A_n})\}_{n \in \mathbb{N}}$

are non-decreasing sequences of real-valued Borel functions on S^2 that converge to $\mathbb{1}_A$ and $\mathcal{L}_\psi(\mathbb{1}_A)$, respectively, as $n \longrightarrow +\infty$. Then by the Lebesgue Monotone Convergence Theorem, we get $A \in \mathfrak{L}$. Hence $\mathfrak{L}$ is a λ-system.

Recall that Dynkin's π-λ theorem (see for example, [Bi95, Theorem 3.2]) states that if $\mathfrak{P}$ is a π-system and $\mathfrak{L}$ is a λ-system that contains $\mathfrak{P}$, then the σ-algebra $\sigma(\mathfrak{P})$ generated by $\mathfrak{P}$ is a subset of $\mathfrak{L}$. Thus by Dynkin's π-λ theorem, the Borel σ-algebra $\sigma(\mathfrak{G})$ is a subset of $\mathfrak{L}$, i.e., equality (3.11) holds for each Borel set $A \subseteq S^2$.

For (ii), we fix a Borel set $A \subseteq S^2$ on which f is injective. By (3.7), we get that $\mathcal{L}_\psi(\mathbb{1}_A)(x) \neq 0$ if and only if $x \in f(A)$. Thus $f(A)$ is Borel. Then (3.10) follows immediately from (i) and (3.7) for $u \in B(S^2)$. □

3.4 Weak Expansion Properties

The topological tail entropy was first introduced by M. Misiurewicz under the name "topological conditional entropy" [Mis73, Mis76]. We adopt the terminology in [Dow11] (see [Dow11, Remark 6.3.18]).

Definition 3.3 (*Topological conditional entropy and topological tail entropy*) Let (X, d) be a compact metric space and $g : X \to X$ a continuous map. The *topological conditional entropy* $h(g|\lambda)$ of g given λ, for some open cover λ, is

$$h(g|\lambda) = \lim_{l\to+\infty} \lim_{n\to+\infty} \frac{1}{n} H\left(\bigvee_{i=0}^{n-1} g^{-i}(\xi_l) \middle| \bigvee_{j=0}^{n-1} g^{-j}(\lambda) \right), \tag{3.13}$$

where $\{\xi_l\}_{l\in\mathbb{N}_0}$ is an arbitrary refining sequence of open covers, and for each pair of open covers ξ and η,

$$H(\xi|\eta) = \log\left(\max_{A\in\eta}\left\{\min\left\{\mathrm{card}\xi_A \,\middle|\, \xi_A \subseteq \xi,\ A \subseteq \bigcup \xi_A\right\}\right\}\right) \tag{3.14}$$

is the logarithm of the minimal number of sets from ξ sufficient to cover any set in η.

The *topological tail entropy* $h^*(g)$ of g is defined by

$$h^*(g) = \lim_{m\to+\infty} \lim_{l\to+\infty} \lim_{n\to+\infty} \frac{1}{n} H\left(\bigvee_{i=0}^{n-1} g^{-i}(\xi_l) \middle| \bigvee_{j=0}^{n-1} g^{-j}(\eta_m) \right), \tag{3.15}$$

where $\{\xi_l\}_{l\in\mathbb{N}_0}$ and $\{\eta_m\}_{m\in\mathbb{N}_0}$ are two arbitrary refining sequences of open covers, and H is as defined in (3.14).

Remark 3.4 The topological entropy of g (see Sect. 3.3) is $h_{\text{top}}(g) = h(g|\{X\})$, where $\{X\}$ is the open cover of X consisting of only one open set X. See for example, [Dow11, Sect. 6.1]. It is also clear from Definition 3.3 that for open covers ξ and η of X, we have $h(g|\xi) \leq h(g|\eta)$ if ξ is a refinement of η.

The limits in (3.13) and (3.15) always exist, and both $h(g|\lambda)$ and $h^*(g)$ are independent of the choices of refining sequences of open covers $\{\xi_l\}_{l\in\mathbb{N}_0}$ and $\{\eta_m\}_{m\in\mathbb{N}_0}$, see [Dow11, Sect. 6.3], especially the comments after [Dow11, Definition 6.3.14].

The topological tail entropy h^* is also well-behaved under iterations, as it satisfies

$$h^*(g^n) = nh^*(g) \tag{3.16}$$

for each $n \in \mathbb{N}$ and each continuous map $g: X \to X$ on a compact metric space X [Mis76, Proposition 3.1].

The concept of h-expansiveness was introduced by R. Bowen in [Bow72]. We adopt the formulation in [Mis76] (see also [Dow11]).

Definition 3.5 (*h-expansiveness*) A continuous map $g : X \to X$ on a compact metric space X is called *h-expansive* if there exists a finite open cover λ of X such that $h(g|\lambda) = 0$.

A weaker property was then introduced by M. Misiurewicz in [Mis73] (see also [Mis76, Dow11]).

Definition 3.6 (*Asymptotic h-expansiveness*) We say that a continuous map $g : X \to X$ on a compact metric space X is *asymptotically h-expansive* if $h^*(g) = 0$.

Recall that a continuous map $g: X \to X$ on a compact metric space X is forward expansive if there exists $\varepsilon > 0$ such that for each $x \in X$, we have $\Phi_\varepsilon(x) = \{x\}$. Here

$$\Phi_\varepsilon(x) = \{y \in X \mid d(g^n(x), g^n(y)) \leq \varepsilon \text{ for all } n \geq 0\},$$

for $\varepsilon > 0$ and $x \in X$. M. Misiurewicz showed that if g is expansive then it is h-expansive, and that if g is and h-expansive then it is asymptotic h-expansive [Mis76]. He also showed that if g is asymptotic h-expansive, then the measure-theoretic entropy $\mu \mapsto h_\mu(g)$ is upper semi-continuous as a function on the space $\mathcal{M}(X, g)$ of g-invariant Borel probability measures equipped with the weak* topology [Mis76].

Chapter 4
The Measure of Maximal Entropy

This chapter is devoted to the study of periodic points and the measure of maximal entropy for an expanding Thurston map.

In Sect. 4.1, we study the fixed points, periodic points, and preperiodic points of the expanding Thurston maps. For the convenience of the reader, we first provide a direct proof in Proposition 4.5, using knowledge from complex dynamics, of the fact that a rational expanding Thurston map R on the Riemann sphere has exactly $1 + \deg R$ fixed points. Then we set out to generalize this result to the class of expanding Thurston maps, and derive the following theorem.

Theorem 4.1 *Every expanding Thurston map* $f : S^2 \to S^2$ *has* $1+\deg f$ *fixed points, counted with weight given by the local degree of the map at each fixed point.*

Here $\deg f$ denotes the topological degree of the map f. The local degree is a natural weight for points on S^2 for expanding Thurston maps.

We first observe that the statement of Theorem 4.1 agrees with what can be concluded from the Lefschetz fixed-point theorem (see for example, [GP10, Chap. 3]) if the map f is smooth and the graph of f intersects the diagonal of $S^2 \times S^2$ transversely at each fixed point of f. However, an expanding Thurston map may not satisfy either of these conditions. It is not clear how to give a proof by using the Lefschetz fixed-point theorem.

The proof of Theorem 4.1 uses the correspondence between the fixed points of f and the 1-tiles in some cell decomposition of S^2 induced by f and its invariant Jordan curve $\mathscr{C} \subseteq S^2$, for the special case when f has a special invariant Jordan curve $\mathscr{C}$. In fact, f may not have such a Jordan curve, but due to the result of [BM17] mentioned above, for each n large enough there exists an f^n-invariant Jordan curve. We use a slightly stronger result as formulated in Lemma 2.17. Then the general case follows from an elementary number-theoretic argument. One of the advantages of this proof is that we also exhibit an almost one-to-one correspondence between the fixed points and the 1-tiles in the cell decomposition of S^2, which leads to precise information

Z. Li, *Ergodic Theory of Expanding Thurston Maps*, Atlantis Studies in Dynamical Systems 4, DOI 10.2991/978-94-6239-174-1_4

on the location of each fixed point. This information is essential later in the proof of the equidistribution of preperiodic and periodic points of expanding Thurston maps in Sect. 4.2. As a corollary of Theorem 4.1, we give a formula in Corollary 4.14 for the number of preperiodic points when counted with the corresponding weight.

In Sect. 4.2, we prove a number of equidistribution results. More precisely, we first prove in Theorem 4.23 the equidistribution of the n-tiles in the tile decompositions discussed in Sect. 2.2 with respect to the measure of maximal entropy μ_f of an expanding Thurston map f. The proof uses a combinatorial characterization of μ_f due to M. Bonk and D. Meyer [BM17] that we will state explicitly in Theorem 4.22.

We then formulate the equidistribution of preimages with respect to the measure of maximal entropy μ_f in Theorem 4.2 below. Here we denote by δ_x the Dirac measure supported on a point x in S^2.

Theorem 4.2 (Equidistribution of preimages) *Let $f : S^2 \to S^2$ be an expanding Thurston map with its measure of maximal entropy μ_f. Fix $p \in S^2$ and define the Borel probability measures*

$$\nu_i = \frac{1}{(\deg f)^i} \sum_{q \in f^{-i}(p)} \deg_{f^i}(q)\delta_q, \qquad \widetilde{\nu}_i = \frac{1}{Z_i} \sum_{q \in f^{-i}(p)} \delta_q, \tag{4.1}$$

for each $i \in \mathbb{N}_0$, where $Z_i = \operatorname{card}\left(f^{-i}(p)\right)$. Then

$$\nu_i \xrightarrow{w^*} \mu_f \text{ as } i \longrightarrow +\infty, \tag{4.2}$$

$$\widetilde{\nu}_i \xrightarrow{w^*} \mu_f \text{ as } i \longrightarrow +\infty. \tag{4.3}$$

Here $\deg_{f^i}(x)$ denotes the local degree of the map f^i at a point $x \in S^2$. In (4.2), (4.3), and similar statements below, the convergence of Borel measures is in the weak* topology, and we use w^* to denote it.

After generalizing Lemma 4.21, which is due to M. Bonk and D. Meyer (see [BM17, Lemma 17.6]), in Lemma 4.28, we prove the equidistribution of preperiodic points with respect to μ_f. Note that Theorem 4.1 is used here.

Theorem 4.3 (Equidistribution of preperiodic points) *Let $f : S^2 \to S^2$ be an expanding Thurston map with its measure of maximal entropy μ_f. For each $m \in \mathbb{N}_0$ and each $n \in \mathbb{N}$ with $m < n$, we define the Borel probability measures*

$$\xi_n^m = \frac{1}{s_n^m} \sum_{f^m(x)=f^n(x)} \deg_{f^n}(x)\delta_x, \qquad \widetilde{\xi}_n^m = \frac{1}{\widetilde{s}_n^m} \sum_{f^m(x)=f^n(x)} \delta_x, \tag{4.4}$$

where $s_n^m, \widetilde{s}_n^m$ are the normalizing factors defined in (4.15) and (4.16). If $\{m_n\}_{n\in\mathbb{N}}$ is a sequence in $\mathbb{N}_0$ such that $m_n < n$ for each $n \in \mathbb{N}$, then

$$\xi_n^{m_n} \xrightarrow{w^*} \mu_f \text{ as } n \longrightarrow +\infty, \tag{4.5}$$

$$\widetilde{\xi}_n^{m_n} \xrightarrow{w^*} \mu_f \text{ as } n \longrightarrow +\infty. \tag{4.6}$$

We prove in Corollary 4.14 that $s_n^m = (\deg f)^n + (\deg f)^m$ for $m \in \mathbb{N}_0$ and $n \in \mathbb{N}$ with $m < n$.

As a special case of Theorem 4.3, we obtain the following corollary.

Corollary 4.4 (Equidistribution of periodic points) *Let $f: S^2 \to S^2$ be an expanding Thurston map with its measure of maximal entropy μ_f. Then*

$$\frac{1}{1 + (\deg f)^n} \sum_{x = f^n(x)} \deg_{f^n}(x) \delta_x \xrightarrow{w^*} \mu_f \text{ as } n \longrightarrow +\infty, \tag{4.7}$$

$$\frac{1}{\operatorname{card}\{x \in S^2 \,|\, x = f^n(x)\}} \sum_{x = f^n(x)} \delta_x \xrightarrow{w^*} \mu_f \text{ as } n \longrightarrow +\infty, \tag{4.8}$$

$$\frac{1}{(\deg f)^n} \sum_{x = f^n(x)} \delta_x \xrightarrow{w^*} \mu_f \text{ as } n \longrightarrow +\infty. \tag{4.9}$$

The equidistribution (4.2), (4.3), (4.7), and (4.8) are analogs of corresponding results for rational maps on the Riemann sphere by Lyubich [Ly83]. Some ideas from [Ly83] are used in the proofs of Theorems 4.2 and 4.3 as well. P. Haïssinsky and K. Pilgrim also proved (4.2) and (4.7) in their general context [HP09], which includes expanding Thurston maps.

The equidistribution (4.5) and (4.6) are inspired by the recent work of M. Baker and L. DeMarco [BD11]. They used some equidistribution result of preperiodic points of rational maps on the Riemann sphere in the context of arithmetic dynamics.

We show in Corollary 4.31 that for each expanding Thurston map f, the exponential growth rate of the cardinality of the set of fixed points of f^n is equal to the topological entropy $h_{\mathrm{top}}(f)$ of f, which is known to be equal to $\log(\deg f)$ (see for example, [BM17, Corollary 17.2]). This is analogous to the corresponding result for expansive homeomorphisms on compact metric spaces with the *specification property* (see for example, [KH95, Theorem 18.5.5]).

In Sect. 4.3, we prove in Theorem 4.33 that for each expanding Thurston map f with its measure of maximal entropy μ_f, the measure-preserving dynamical system (S^2, f, μ_f) is a factor, in the category of measure-preserving dynamical systems, of the measure-preserving dynamical system of the left-shift operator on the one-sided infinite sequences of $\deg f$ symbols together with its measure of maximal entropy. This generalizes the corresponding result in [BM17] in the category of topological dynamical systems, reformulated in Theorem 4.32.

4.1 Number and Locations of Fixed Points

The main goal of this section is to prove Theorem 4.1; namely, that the number of fixed points, counted with appropriate weights, of an expanding Thurston map f is exactly $1+\deg f$. In order to prove Theorem 4.1, we first establish in Lemmas 4.6 and 4.7 an almost one-to-one correspondence between fixed points and 1-tiles in the cell decomposition $\mathbf{D}^1(f, \mathscr{C})$ for an expanding Thurston map f with an f-invariant Jordan curve $\mathscr{C}$ containing post f. As a consequence, we establish in Corollary 4.14 an exact formula for the number of preperiodic points, counted with appropriate weights. We end this section by establishing a formula for the exact number of periodic points with period n, $n \in \mathbb{N}$, for expanding Thurston maps without periodic critical points.

Let f be a Thurston map and $p \in S^2$ a periodic point of f of period $n \in \mathbb{N}$, we define the *weight of p (with respect to f)* as the local degree $\deg_{f^n}(p)$ of f^n at p. When f is understood from the context and p is a fixed point of f, we abbreviate it as the *weight of p*. We will prove in this section that each expanding Thurston map f has exactly $1 + \deg f$ fixed points, counted with weight.

Note the difference between the weight and the *multiplicity* of a fixed point of a rational map (see [Mil06, Chap. 12]). In comparison, the multiplicity of a fixed point $p \in \mathbb{C}$ of a rational map $g : \widehat{\mathbb{C}} \to \widehat{\mathbb{C}}$ is $\deg_{\widetilde{g}}(p)$, where $\widetilde{g}(z) = g(z) - z$. For every expanding rational Thurston map $R: \widehat{\mathbb{C}} \to \widehat{\mathbb{C}}$, M. Bonk and D. Meyer proved that R has no periodic critical points (see [BM17, Proposition 2.3]). So the weight of every fixed point of R is 1. We can prove that R has exactly $1 + \deg R$ fixed points by using basic facts in complex dynamics, even though it will follow as a special case of our general result in Theorem 4.1. For the relevant definitions and general background of complex dynamics, see [CG93, Mil06].

Proposition 4.5 *Let $R: \widehat{\mathbb{C}} \to \widehat{\mathbb{C}}$ be a expanding rational Thurston map, then R has exactly $1 + \deg R$ fixed points. Moreover, the weight $\deg_R(q)$ of each fixed point q of R is equal to 1.*

Proof Conjugating R by a fractional linear automorphism of the Riemann sphere if necessary, we may assume that the point at infinity is not a fixed point of R.

Since R is expanding, R is not the identity map. By Lemma 12.1 in [Mil06], which is basically an application of the fundamental theorem of algebra, we can conclude that R has $1 + \deg R$ fixed points, counted with multiplicity. For rational Thurston maps, being expanding is equivalent to having no periodic critical points (see [BM17, Proposition 2.3]). So the weight $\deg_R(q)$ of every fixed point q of R is exactly 1. Thus it suffices now to prove that each fixed point q of R has multiplicity 1.

Suppose a fixed point p of R has multiplicity $m > 1$. In the terminology of complex dynamics, q is then a parabolic fixed point with multiplier 1 and multiplicity m. Then by Leau-Fatou flower theorem (see for example, [Mil06, Chap. 10] or [Br10, Theorem 2.12]), there exists an open set $U \subseteq S^2$ such that $f(U) \subseteq U$ and $U \neq S^2$ (by letting U be one of the attracting petals, for example). This contradicts the fact that the function R, as an expanding Thurston map, is *eventually onto*, i.e., for each nonempty open set $V \subseteq S^2$, there exists a number $m \in \mathbb{N}$ such that $R^m(V) = S^2$.

In order to see that R is eventually onto, let d be a metric on S^2 and $\mathscr{C} \subseteq S^2$ be a Jordan curve, as given in Definition 2.10. Since V is open, it contains some open ball in the metric space (S^2, d). Then since R is expanding, by Definition 2.10, we can conclude that there exist a constant $m \in \mathbb{N}$, a black m-tile $X_b^m \in \mathbf{X}_b^m(R, \mathscr{C})$ and a white m-tile $X_w^m \in \mathbf{X}_w^m(R, \mathscr{C})$ such that $X_b^m \cup X_w^m \subseteq V$. Thus $R^m(V) \supseteq R^m(X_b^m \cup X_w^m) = S^2$. Therefore, R is eventually onto. □

For general expanding Thurston maps, we need to use the combinatorial information from [BM17].

Lemma 4.6 *Let f be an expanding Thurston map with an f-invariant Jordan curve $\mathscr{C}$ containing* post f. *If $X \in \mathbf{X}_{ww}^1(f, \mathscr{C}) \cup \mathbf{X}_{bb}^1(f, \mathscr{C})$ is a white 1-tile contained in the while 0-tile X_w^0 or a black 1-tile contained in the black 0-tile X_b^0, then X contains at least one fixed point of f. If $X \in \mathbf{X}_{wb}^1(f, \mathscr{C}) \cup \mathbf{X}_{bw}^1(f, \mathscr{C})$ is a white 1-tile contained in the black 0-tile X_b^0 or a black 1-tile contained in the white 0-tile X_w^0, then* inte(X) *contains no fixed points of f.*

Recall that cells in the cell decompositions are by definition closed sets, and the set of 0-tiles $\mathbf{X}^0(f, \mathscr{C})$ consists of the white 0-tile X_w^0 and the black 0-tile X_b^0.

Proof If $X \in \mathbf{X}_{ww}^1(f, \mathscr{C}) \cup \mathbf{X}_{bb}^1(f, \mathscr{C})$, then $X \subseteq f(X)$. By Proposition 2.6(i), $f|_X$ is a homeomorphism from X to $f(X)$, which is one of the two 0-tiles. Hence, $f(X)$ is homeomorphic to the closed unit disk. So by Brouwer's fixed point theorem, $(f|_X)^{-1}$ has a fixed point p. Thus p is also a fixed point of f.

If $X \in \mathbf{X}_{wb}^1(f, \mathscr{C})$, then $\text{inte}(X) \subseteq \text{inte}(X_b^0)$ and $f(X) = X_w^0$. Since $X_w^0 \cap \text{inte}(X_b^0) = \emptyset$, the map f has no fixed points in inte(X). The case when $X \in \mathbf{X}_{bw}^1(f, \mathscr{C})$ is similar. □

Lemma 4.7 *Let f be an expanding Thurston map with an f-invariant Jordan curve $\mathscr{C}$ containing* post f *such that no 1-tile in $\mathbf{D}^1(f, \mathscr{C})$ joins opposite sides of $\mathscr{C}$. Then for every $n \in \mathbb{N}$, each n-tile $X^n \in \mathbf{X}^n(f, \mathscr{C})$ contains at most one fixed point of f^n.*

Proof Fix an arbitrary $n \in \mathbb{N}$. We denote $F = f^n$ and consider the cell decompositions induced by F and $\mathscr{C}$ in this proof. Note that F is also an expanding Thurston map and there is no 1-tile in $\mathbf{D}^1(F, \mathscr{C})$ joining opposite sides of $\mathscr{C}$.

It suffices to prove that each 1-tile $X^1 \in \mathbf{X}^1$ contains at most one fixed point of F.

Suppose that there are two distinct fixed points p, q of F in a 1-tile X^1. We prove that there is a contradiction in each of the following cases.

Case 1: at least one of the fixed points, say p, is in inte(X^1). Then $X^1 \in \mathbf{X}_{ww}^1 \cup \mathbf{X}_{bb}^1$ by Lemma 4.6. Since p is contained in the interior of $X_1 \cap F(X_1)$, we get that $X_1 \subseteq F(X_1)$. Since $F|_{X^1}$ is a homeomorphism from X^1 to $F(X^1)$ (see Proposition 2.6(i)), we define a 2-tile $X^2 = (F|_{X^1})^{-1}(X^1) \subseteq X^1$. Then we get that $p \in \text{inte}(X^2)$ and $F(X^2) = X^1$. On the other hand, the point q must be in X^2 as well for otherwise there exists $q' \neq q$ such that $q' \in X^2$ and $F(q') = q$, thus q' and q are two distinct points in X^1 whose images under F are q, contradicting the fact that $F|_{X^1}$ is a homeomorphism from X^1 to $F(X^1)$ and $X^1 \subseteq F(X^1)$. Similarly we can inductively construct an $(n + 1)$-cell

$X^{n+1} \subseteq X^n$ such that $F(X^{n+1}) = X^n$, $p \in \text{inte}(X^{n+1})$, and $q \in X^{n+1}$, for each $n \in \mathbb{N}$. This contradicts the fact that F is an expanding Thurston map, see Remark 2.11.

Case 2: there exists a 1-edge $e \in \mathbf{E}^1$ such that $p, q \in e$. Note that $e \subseteq X^1$. Then one of the fixed points p and q, say p, must be contained in the interior of e, for otherwise p, q are distinct 1-vertices that are fixed by F, thus they are both 0-vertices, hence X^1 joins opposite sides, a contradiction. Since $F(e)$ is a 0-edge by Proposition 2.6, and $p \in F(e)$, there exists a 1-edge $e' \subseteq F(e)$ with $p \in e'$. Thus e' intersects with e at the point p, which is an interior point of e. So $e' = e$, and $e \subseteq F(e)$. Then by the same argument as when $p \in \text{inte}(X^1)$ in Case 1, we can get a contradiction to the fact that F is an expanding Thurston map.

Case 3: the points p, q are contained in two distinct 1-edges e_1, e_2 of X^1, respectively, and $e_1 \cap e_2 \neq \emptyset$. Since F is an expanding Thurston map, we have $m = \text{card}(\text{post } F) \geq 3$ (see [BM17, Corollary 7.2]). So X^1 is an m-gon (see Proposition 2.6(vi)). Since $e_1 \cap e_2 \neq \emptyset$, we get $\text{card}(e_1 \cap e_2) = 1$, say $e_1 \cap e_2 = \{v\}$. By Case 2, we get that $v \neq p$ and $v \neq q$. Note that $p \in F(e_1)$, $q \in F(e_2)$, and $F(e_1), F(e_2)$ are 0-edges. If at least one of p and q is a 1-vertex, thus a 0-vertex as well, then since Proposition 2.6(i) implies that $F(e_1) \neq F(e_2)$, we can conclude that X^1 touches at least three 0-edges, thus joins opposite sides of $\mathscr{C}$, a contradiction. Hence $p \in \text{inte}(e_1)$ and $q \in \text{inte}(e_2)$. So $e_1 \subseteq F(e_1)$, $e_2 \subseteq F(e_2)$, and

$$\{v\} = e_1 \cap e_2 \subseteq F(e_1) \cap F(e_2) = F(e_1 \cap e_2) = F(\{v\}),$$

by Proposition 2.6(i). Thus $F(v) = v$. Then e_1 contains two distinct fixed points p and v of F, which is impossible by Case 2.

Case 4: the points p, q are contained in two distinct 1-edges e_1, e_2 of X^1, respectively, and $e_1 \cap e_2 = \emptyset$. Thus $\text{card}(\text{post } F) \geq 4$ by Proposition 2.6(vi), and $F(e_1)$ and $F(e_2)$ are a pair of disjoint edges of $F(X^1)$ by Proposition 2.6(i). But $p = F(p) \in F(e_1)$, $q = F(q) \in F(e_2)$, so X^1 joins opposite sides of $\mathscr{C}$, a contradiction.

Combining all cases above, we can conclude, therefore, that each 1-tile $X^1 \in \mathbf{X}^1$ contains at most one fixed point of F. □

We can immediately get an upper bound of the number of periodic points of an expanding Thurston map from Lemma 4.7.

Corollary 4.8 *Let f be an expanding Thurston map. Then for each $n \in \mathbb{N}$ sufficiently large, the number of fixed points of f^n is $\leq 2(\deg f)^n$. In particular, the number of fixed points of f is finite.*

Proof By Corollary 2.18, for each $n \geq N(f)$, where $N(f) \in \mathbb{N}$ is a constant as given in Corollary 2.18, there exists an f^n-invariant Jordan curve $\mathscr{C}$ containing post f such that no n-tile in $\mathbf{D}^n(f, \mathscr{C})$ joins opposite sides of $\mathscr{C}$. Let $F = f^n$. So F is an expanding Thurston map, and $\mathscr{C}$ is an F-invariant Jordan curve containing post F such that no 1-tile in $\mathbf{D}^1(F, \mathscr{C})$ joins opposite sides of $\mathscr{C}$. By Proposition 2.6(iv), the number of

1-tiles in $\mathbf{X}^1(F, \mathscr{C})$ is exactly $2 \deg F = 2(\deg f)^n$. By Lemma 4.7, we can conclude that there are at most $2(\deg f)^n$ fixed points of $F = f^n$.

Since each fixed point of f is also a fixed point of f^n, for each $n \in \mathbb{N}$, the number of fixed points of f is finite. □

The following lemma in some sense generalizes Lemma 4.7 to Jordan curves that are not necessarily f-invariant, but f^{n_c}-invariant for some $n_c \in \mathbb{N}$. The conclusions of both lemmas hold when n is sufficiently large, which is a combinatorial condition in Lemma 4.7 and a metric condition for the following lemma. The proof of the following lemma is simpler, but the proof of Lemma 4.7 is more self-contained. The following lemma will not be used until Chap. 7.

Lemma 4.9 *Let $f : S^2 \to S^2$ be an expanding Thurston map, and $\mathscr{C} \subseteq S^2$ be a Jordan curve that satisfies* $\operatorname{post} f \subseteq \mathscr{C}$ *and $f^{n_{\mathscr{C}}}(\mathscr{C}) \subseteq \mathscr{C}$ for some $n_{\mathscr{C}} \in \mathbb{N}$. Let d be a visual metric on S^2 for f with expansion factor $\Lambda > 1$. Then there exists $N_0 \in \mathbb{N}$ such that for each $n \geq N_0$ and each n-tile $X^n \in \mathbf{X}^n(f, \mathscr{C})$, the number of fixed points of f^n contained in X^n is at most* 1.

Proof By Lemma 2.19, for each $i \in \mathbb{N}$, each i-tile $X^i \in \mathbf{X}^i(f, \mathscr{C})$, and each pair of points $x, y \in X^i$, we have

$$d\left(f^i(x), f^i(y)\right) \geq \frac{\Lambda^i}{C_0} d(x, y),$$

where $C_0 > 1$ is a constant depending only on f and d from Lemma 2.19.

We choose $N_0 \in \mathbb{N}$ such that $\Lambda^{N_0} > C_0$.

Let $n \geq N_0$ and $X^n \in \mathbf{X}^n(f, \mathscr{C})$. Suppose two distinct points $p, q \in X$ satisfy $f^n(p) = p$ and $f^n(q) = q$. Then

$$1 = \frac{d\left(f^n(p), f^n(q)\right)}{d(p, q)} \geq \frac{\Lambda^n}{C_0} > 1,$$

a contradiction. This completes the proof. □

Lemma 4.10 *Let f be an expanding Thurston map with an f-invariant Jordan curve $\mathscr{C}$ containing* $\operatorname{post} f$. *Then*

$$\begin{aligned} \deg(f|_{\mathscr{C}}) &= \operatorname{card}\left(\mathbf{X}^1_{ww}(f, \mathscr{C})\right) - \operatorname{card}\left(\mathbf{X}^1_{bw}(f, \mathscr{C})\right) \\ &= \operatorname{card}\left(\mathbf{X}^1_{bb}(f, \mathscr{C})\right) - \operatorname{card}\left(\mathbf{X}^1_{wb}(f, \mathscr{C})\right). \end{aligned} \tag{4.10}$$

Here $\deg(f|_{\mathscr{C}})$ is the *degree* of the map $f|_{\mathscr{C}} : \mathscr{C} \to \mathscr{C}$. Roughly speaking, it measures the total number of times the image of $\mathscr{C}$ under f winds around $\mathscr{C}$ along the orientation of $\mathscr{C}$. See for example, [Ha02, Sect. 2.2] for a precise definition.

Note that the first equality in (4.10), for example, says that the degree of f restricted to $\mathscr{C}$ is equal to the number of white 1-tiles contained in the white 0-tile minus the number of black 1-tiles contained in the white 0-tile.

Recall that for each continuous path $\gamma\colon [a,b] \to \mathbb{C}\backslash\{0\}$ on the Riemann sphere $\widehat{\mathbb{C}}$, with $a, b \in \mathbb{R}$ and $a < b$, we can define the *variation of the argument along* γ, denoted by $V(\gamma)$, as the change of the imaginary part of the logarithm along γ. Note that $V(\gamma)$ is invariant under an orientation-preserving reparametrization of γ and if $\widetilde{\gamma}\colon [a,b] \to \widehat{\mathbb{C}}$ reverses the orientation of γ, i.e., $\widetilde{\gamma}(t) = \gamma(a+b-t)$, then $V(\widetilde{\gamma}) = -V(\gamma)$. We also note that if γ is a loop, then $V(\gamma) = 2\pi \operatorname{Ind}_\gamma(0)$, where $\operatorname{Ind}_\gamma(0)$ is the *winding number of* γ *with respect to* 0 [Burc79, Chap. IV].

Proof Consider the cell decompositions induced by $(f, \mathscr{C})$. Let X^0_w be the white 0-tile.

We start with proving the first equality in (4.10).

By the Schoenflies theorem (see for example, [Moi77, Theorem 10.4]), we can assume that S^2 is the Riemann sphere $\widehat{\mathbb{C}}$, and X^0_w is the unit disk with the center 0 disjoint from $f^{-1}(\mathscr{C})$.

For each 1-edge $e \in \mathbf{E}^1$, we choose a parametrization $\gamma_e^+\colon [0,1] \to \mathbb{C}\backslash\{0\}$ of e with positive orientation (i.e., with the white 1-tile on the left), and a parametrization $\gamma_e^-\colon [0,1] \to \mathbb{C}\backslash\{0\}$ of e with negative orientation. Then $f \circ \gamma_e^+$ and $f \circ \gamma_e^-$ are parametrizations of one of the 0-edges on the unit circle $\mathscr{C}$, with positive orientation and negative orientation, respectively.

We claim that

$$\sum_{X\in\mathbf{X}^1_{ww}} \sum_{e\in\mathbf{E}^1, e\subseteq\partial X} V(f\circ\gamma_e^+) - \sum_{X\in\mathbf{X}^1_{bw}} \sum_{e\in\mathbf{E}^1, e\subseteq\partial X} V(f\circ\gamma_e^+) = \sum_{e\in\mathbf{E}^1, e\subseteq\mathscr{C}} V(f\circ\gamma_e), \quad (4.11)$$

where on the right-hand side, $\gamma_e = \gamma_e^+$ if $e \subseteq \mathscr{C} \cap X$ for some $X \in \mathbf{X}^1_{ww}$ and $\gamma_e = \gamma_e^-$ if $e \subseteq \mathscr{C} \cap X$ for some $X \in \mathbf{X}^1_{bw}$, or equivalently, γ_e parametrizes e in such a way that X^0_w is always on the left of e for each $e \in \mathbf{E}^1$ with $e \subseteq \mathscr{C}$.

We observe that the left-hand side of (4.11) is the sum of $V(f \circ \gamma_e^+)$ over all 1-edges e in the boundary of a white 1-tile $X \subseteq X^0_w$ plus the sum of $V(f \circ \gamma_e^-)$ over all 1-edges e in the boundary of a black 1-tile $X \subseteq X^0_w$. Since each 1-edge e with $\operatorname{inte}(e) \subseteq X^0_w$ is the intersection of exactly one 1-tile in $\mathbf{X}^1_{ww}$ and one 1-tile in $\mathbf{X}^1_{bw}$, the two terms corresponding to a 1-edge e that is not contained in $\mathscr{C}$ cancel each other. Moreover, there is exactly one term for each 1-edge $e \subseteq X^0_w$ that is contained in $\mathscr{C}$, and e that corresponds to such a term is parametrized in such a way that X^0_w is on the left of e. The claim now follows.

We then note that by Proposition 2.6(i), the left-hand side of (4.11) is equal to

$$\sum_{X\in\mathbf{X}^1_{ww}} 2\pi - \sum_{X\in\mathbf{X}^1_{bw}} 2\pi = 2\pi \left(\operatorname{card}(\mathbf{X}^1_{ww}) - \operatorname{card}(\mathbf{X}^1_{bw})\right),$$

and the right-hand side of (4.11) is equal to

$$2\pi \operatorname{Ind}_{f\circ\gamma_{\mathscr{C}}}(0) = 2\pi \deg(f|_{\mathscr{C}}),$$

where $\gamma_{\mathscr{C}}$ is a parametrization of $\mathscr{C}$ with positive orientation. Hence the first equality in (4.10) follows.

The second equality in (4.10) follows by symmetry, in the sense that we could have exchanged the colors of the 0-tiles and thus exchanged the colors of all tiles. It also follows from the fact that

$$\operatorname{card}(\mathbf{X}^1_{ww}) + \operatorname{card}(\mathbf{X}^1_{wb}) = \deg f = \operatorname{card}(\mathbf{X}^1_{bb}) + \operatorname{card}(\mathbf{X}^1_{bw}).$$

□

Let f be an expanding Thurston map with an f-invariant Jordan curve $\mathscr{C}$ containing post f. We orient $\mathscr{C}$ in such a way that the white 0-tile lies on the left of $\mathscr{C}$. Let $p \in \mathscr{C}$ be a fixed point of f. We say that $f|_{\mathscr{C}}$ *preserves the orientation at* p (resp. *reverses the orientation at* p) if there exists an open arc $l \subseteq \mathscr{C}$ with $p \in l$ such that f maps l homeomorphically to $f(l)$ and $f|_{\mathscr{C}}$ preserves (resp. reverses) the orientation on l. More concretely, when p is a 1-vertex, let $l_1, l_2 \subseteq \mathscr{C}$ be the two distinct 1-edges on $\mathscr{C}$ containing p; when $p \in \operatorname{inte}(e)$ for some 1-edge $e \subseteq \mathscr{C}$, let l_1, l_2 be the two connected components of $e \setminus \{p\}$. Then $f|_{\mathscr{C}}$ preserves the orientation at p if $l_1 \subseteq f(l_1)$ and $l_2 \subseteq f(l_2)$, and reverses the orientation at p if $l_2 \subseteq f(l_1)$ and $l_1 \subseteq f(l_2)$. Note that it may happen that $f|_{\mathscr{C}}$ neither preserves nor reverses the orientation at p, because $f|_{\mathscr{C}}$ need not be a local homeomorphism near p, where it may behave like a "folding map".

Lemma 4.11 *Let f be an expanding Thurston map with an f-invariant Jordan curve $\mathscr{C}$ containing* post f. *Then the number of fixed points of $f|_{\mathscr{C}}$ where $f|_{\mathscr{C}}$ preserves the orientation minus the number of fixed points of $f|_{\mathscr{C}}$ where $f|_{\mathscr{C}}$ reverses the orientation is equal to* $\deg(f|_{\mathscr{C}}) - 1$.

Proof Let $\psi\colon [0,1] \to \mathscr{C}$ be a continuous map such that $\psi|_{(0,1)}\colon (0,1) \to \mathscr{C}\setminus\{x_0\}$ is an orientation-preserving homeomorphism, and $\psi(0) = \psi(1) = x_0$ for some $x_0 \in \mathscr{C}$ that is not a fixed point of $f|_{\mathscr{C}}$. Note that for each $x \in \mathscr{C}$ with $x \neq x_0$, $\psi^{-1}(x)$ is a well-defined number in $(0,1)$. In particular, $\psi^{-1}(y)$ is a well-defined number in $(0,1)$ for each fixed point y of $f|_{\mathscr{C}}$. Define $\pi\colon \mathbb{R} \to \mathscr{C}$ by $\pi(x) = \psi(x - \lfloor x \rfloor)$. Then π is a covering map. We lift $f|_{\mathscr{C}} \circ \psi$ to $G\colon [0,1] \to \mathbb{R}$ such that $\pi \circ G = f|_{\mathscr{C}} \circ \psi$ and $G(0) = \psi^{-1}(f(x_0)) \in (0,1)$. So we get the following commutative diagram:

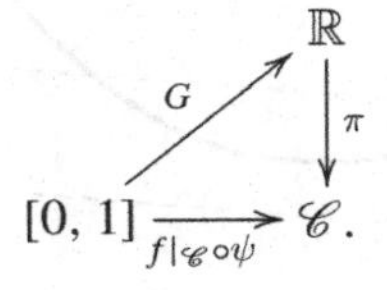

Then $G(1) - G(0) \in \mathbb{Z}$ and

$$\deg(f|_{\mathscr{C}}) = G(1) - G(0). \tag{4.12}$$

Observe that $y \in \mathscr{C}$ is a fixed point of $f|_{\mathscr{C}}$ if and only if $G(\psi^{-1}(y)) - \psi^{-1}(y) \in \mathbb{Z}$. Indeed, if $y \in \mathscr{C}$ is a fixed point of $f|_{\mathscr{C}}$, then $\pi \circ G \circ \psi^{-1}(y) = f|_{\mathscr{C}}(y) = y$. Thus $G \circ \psi^{-1}(y) - \psi^{-1}(y) \in \mathbb{Z}$. Conversely, if $G \circ \psi^{-1}(y) - \psi^{-1}(y) \in \mathbb{Z}$, then $y \neq x_0$ since $G(\psi^{-1}(x_0)) - \psi^{-1}(x_0) = G(0) - 0 \in (0, 1)$, thus

$$f|_{\mathscr{C}}(y) = f|_{\mathscr{C}} \circ \psi \circ \psi^{-1}(y) = \pi \circ G \circ \psi^{-1}(y) = \pi \circ \psi^{-1}(y) = y.$$

For each $m \in \mathbb{Z}$, we define the line l_m to be the graph of the function $x \mapsto x + m$ from $\mathbb{R}$ to $\mathbb{R}$.

Let $y \in \mathscr{C}$ be any fixed point of $f|_{\mathscr{C}}$. Since by Corollary 4.8 fixed points of f are isolated, there exists a neighborhood $(s, t) \subseteq (0, 1)$ such that $\psi^{-1}(y) \in (s, t)$ and for each fixed point $z \in \mathscr{C}\backslash\{y\}$ of $f|_{\mathscr{C}}$, $\psi^{-1}(z) \notin (s, t)$. Define $k = G(\psi^{-1}(y)) - \psi^{-1}(y)$; then $k \in \mathbb{Z}$. Moreover, $z \in \mathscr{C}$ is a fixed point of $f|_{\mathscr{C}}$ if and only if the graph of G intersects with l_m at the point $\left(\psi^{-1}(z), G\left(\psi^{-1}(z)\right)\right)$ for some $m \in \mathbb{Z}$.

Depending on the orientation of $f|_{\mathscr{C}}$ at the fixed point $y \in \mathscr{C}$, we get one of the following cases (see for example, Fig. 4.1):

1. If $f|_{\mathscr{C}}$ preserves the orientation at y, then the graph of $G|_{(s,\psi^{-1}(y))}$ lies strictly between the lines l_{k-1} and l_k, and the graph of $G|_{(\psi^{-1}(y),t)}$ lies strictly between the lines l_k and l_{k+1}.
2. If $f|_{\mathscr{C}}$ reverses the orientation at y, then the graph of $G|_{(s,\psi^{-1}(y))}$ lies strictly between the lines l_k and l_{k+1}, and the graph of $G|_{(\psi^{-1}(y),t)}$ lies strictly between the lines l_{k-1} and l_k.
3. If $f|_{\mathscr{C}}$ neither preserves nor reverses the orientation at y, then the graph of $G|_{(s,t)\backslash\{\psi^{-1}(y)\}}$ either lies strictly between the lines l_{k-1} and l_k or lies strictly between the lines l_k and l_{k+1}.

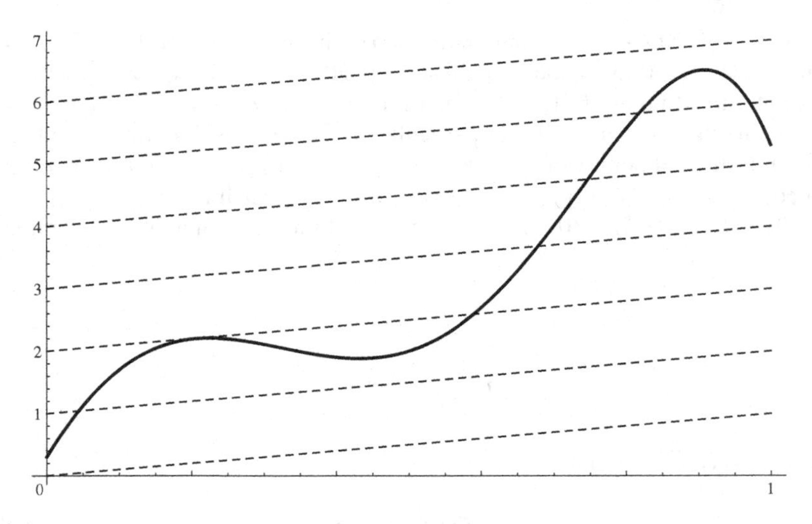

Fig. 4.1 The lines l_k for $k \in \mathbb{Z}$ and an example of the graph of G

Thus the number of fixed points of $f|_{\mathscr{C}}$ where $f|_{\mathscr{C}}$ preserves the orientation is exactly the number of intersections between the graph of G and the lines l_m with $m \in \mathbb{Z}$, where the graph of G crosses the lines from below, and the number of fixed points of $f|_{\mathscr{C}}$ where $f|_{\mathscr{C}}$ reverses the orientation is exactly the number of intersections between the graph of G and the lines l_m with $m \in \mathbb{Z}$, where the graph of G crosses the lines from above. Therefore the number of fixed points of $f|_{\mathscr{C}}$ where $f|_{\mathscr{C}}$ preserves the orientation minus the number of fixed points of $f|_{\mathscr{C}}$ where $f|_{\mathscr{C}}$ reverses the orientation is equal to $G(1) - G(0) - 1 = \deg(f|_{\mathscr{C}}) - 1$. □

For each $n \in \mathbb{N}$ and each expanding Thurston map $f: S^2 \to S^2$, we denote by

$$P_{n,f} = \{x \in S^2 \,|\, f^n(x) = x, f^k(x) \neq x, k \in \{1, 2, \ldots, n-1\}\} \tag{4.13}$$

the *set of periodic points of f with period n*, and by

$$p_{n,f} = \sum_{x \in P_{n,f}} \deg_{f^n}(x), \qquad \widetilde{p}_{n,f} = \operatorname{card} P_{n,f} \tag{4.14}$$

the numbers of periodic points x of f with period n, counted with and without weight $\deg_{f^n}(x)$, respectively, at each x. In particular, $P_{1,f}$ is the set of fixed points of f and $p_{1,f} = 1 + \deg f$ as we will see in the proof of Theorem 4.1 below. More generally, for all $m \in \mathbb{N}_0$ and $n \in \mathbb{N}$ with $m < n$, we denote by

$$S_n^m = \{x \in S^2 \,|\, f^m(x) = f^n(x)\} \tag{4.15}$$

the *set of preperiodic points of f with parameters m, n* and by

$$s_n^m = \sum_{x \in S_n^m} \deg_{f^n}(x), \qquad \widetilde{s}_n^m = \operatorname{card} S_n^m \tag{4.16}$$

the numbers of preperiodic points of f with parameters m, n, counted with and without weight $\deg_{f^n}(x)$, respectively, at each x. Note that in particular, for each $n \in \mathbb{N}$, $S_n^0 = P_{1,f^n}$ is the set of fixed points of f^n.

Proposition 4.12 *Let $F: S^2 \to S^2$ be an expanding Thurston map with an F-invariant Jordan curve $\mathscr{C}$ containing* post F *such that no 1-tile in $\mathbf{D}^1(F, \mathscr{C})$ joins opposite sides of $\mathscr{C}$. Then F has $1 + \deg F$ fixed points, counted with weight given by the local degree $\deg_F(x)$ of the map at each fixed point x.*

Proof We consider the cell decompositions induced by $(F, \mathscr{C})$ in this proof. Let $w_w = \operatorname{card} \mathbf{X}^1_{ww}$ be the number of white 1-tiles contained in the white 0-tile, $b_w = \operatorname{card} \mathbf{X}^1_{bw}$ be the number of black 1-tiles contained in the white 0-tile, $w_b = \operatorname{card} \mathbf{X}^1_{wb}$ be the number of white 1-tiles contained in the black 0-tile, and $b_b = \operatorname{card} \mathbf{X}^1_{bb}$ be the number of black 1-tiles contained in the black 0-tile. Note that $w_w + w_b = b_w + b_b = \deg F$.

By Corollary 4.8, we know that fixed points of F are isolated.

Note that

$$w_w + b_b = \deg F + \deg(F|_{\mathscr{C}}), \tag{4.17}$$

which follows from the equation $w_w - b_w = \deg(F|_{\mathscr{C}})$ by Lemma 4.10, and the equation $b_w + b_b = \deg F$.

We define sets

$$A = \{X \in \mathbf{X}^1_{ww} \mid \text{there exists}\, p \in \mathscr{C} \cap X \,\text{with}\, F(p) = p\},$$

$$B = \{X \in \mathbf{X}^1_{bw} \mid \text{there exists}\, p \in \mathscr{C} \cap X \,\text{with}\, F(p) = p\},$$

and let $a = \operatorname{card} A$, $b = \operatorname{card} B$.

We then claim that

$$a - b = \deg(F|_{\mathscr{C}}) - 1. \tag{4.18}$$

In order to prove this claim, we will first prove that $a - b$ is equal to the number of fixed points of $F|_{\mathscr{C}}$ where $F|_{\mathscr{C}}$ preserves the orientation minus the number of fixed points of $F|_{\mathscr{C}}$ where $F|_{\mathscr{C}}$ reverses the orientation.

So let $p \in \mathscr{C}$ be a fixed point of $F|_{\mathscr{C}}$.

1. If p is not a critical point of F, then either $F|_{\mathscr{C}}$ preserves or reverses the orientation at p. In this case, the point p is contained in exactly one white 1-tile and one black 1-tile.
   a. If $F|_{\mathscr{C}}$ preserves the orientation at p, then p is contained in exactly one white 1-tile that is contained in the white 0-tile, and p is not contained in any black 1-tile that is contained in the while 0-tile.
   b. If $F|_{\mathscr{C}}$ reverses the orientation at p, then p is contained in exactly one black 1-tile that is contained in the white 0-tile, and p is not contained in any white 1-tile that is contained in the while 0-tile.
2. If p is a critical point of F, then $p = F(p) \in \operatorname{post} f$ and so there are two distinct 1-edges $e_1, e_2 \subseteq \mathscr{C}$ such that $\{p\} = e_1 \cap e_2$. We refer to Figs. 4.2, 4.3, 4.4 and 4.5.
   a. If $e_1 \subseteq F(e_1)$ and $e_2 \subseteq F(e_2)$, then p is contained in exactly k white and $k-1$ black 1-tiles that are contained in the white 0-tile, for some $k \in \mathbb{N}$. Note that in this case $F|_{\mathscr{C}}$ preserves the orientation at p.
   b. If $e_2 \subseteq F(e_1)$ and $e_1 \subseteq F(e_2)$, then p is contained in exactly $k-1$ white and k black 1-tiles that are contained in the white 0-tile, for some $k \in \mathbb{N}$. Note that in this case $F|_{\mathscr{C}}$ reverses the orientation at p.
   c. If $e_1 \subseteq F(e_1) = F(e_2)$, then p is contained in exactly k white and k black 1-tiles that are contained in the white 0-tile, for some $k \in \mathbb{N}$. Note that in this case $F|_{\mathscr{C}}$ neither preserves nor reverses the orientation at p.
   d. If $e_2 \subseteq F(e_1) = F(e_2)$, then p is contained in exactly k white and k black 1-tiles that are contained in the white 0-tile, for some $k \in \mathbb{N}$. Note that in this case $F|_{\mathscr{C}}$ neither preserves nor reverses the orientation at p.

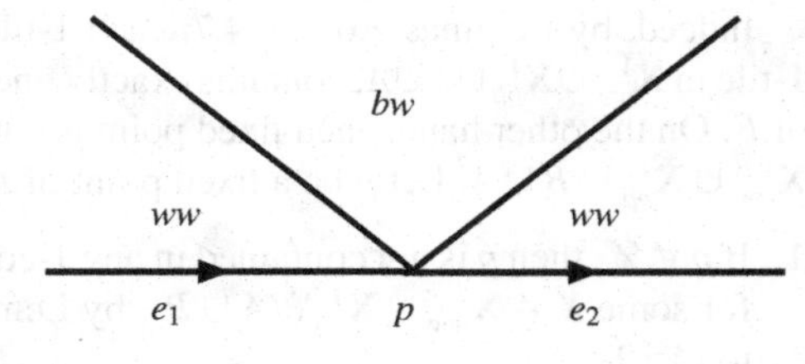

Fig. 4.2 Case (2)(a) where $F(e_1) \supseteq e_1$ and $F(e_2) \supseteq e_2$

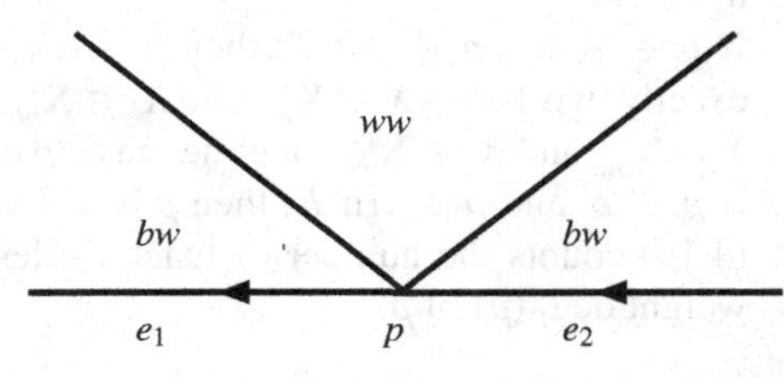

Fig. 4.3 Case (2)(b) where $F(e_1) \supseteq e_2$ and $F(e_2) \supseteq e_1$

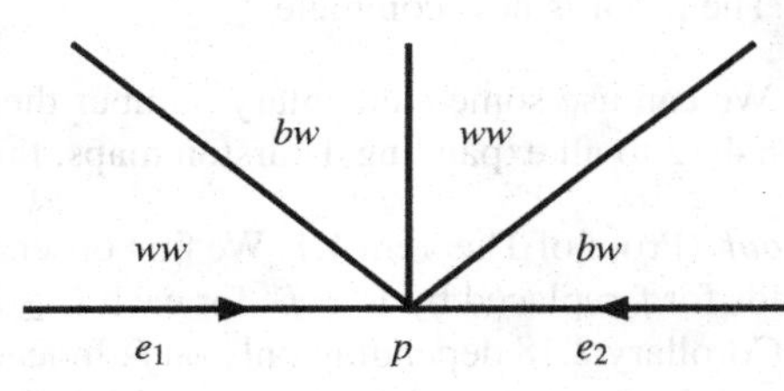

Fig. 4.4 Case (2)(c) where $F(e_1) = F(e_2) \supseteq e_1$

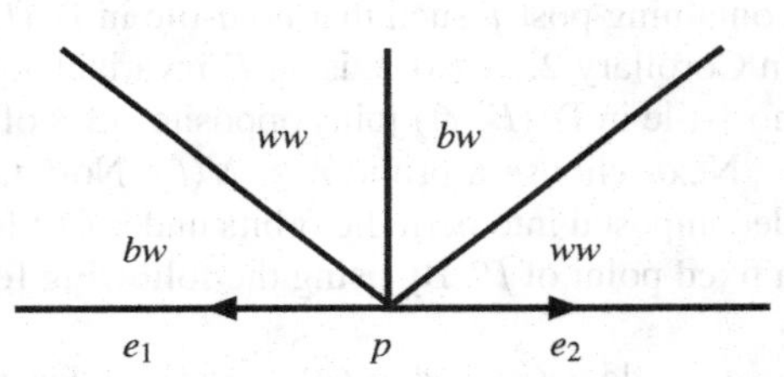

Fig. 4.5 Case (2)(d) where $F(e_1) = F(e_2) \supseteq e_2$

It follows then that $a - b$ is equal to the number of fixed points of $F|_{\mathscr{C}}$ where $F|_{\mathscr{C}}$ preserves the orientation minus the number of fixed points of $F|_{\mathscr{C}}$ where $F|_{\mathscr{C}}$ reverses the orientation.

Then the claim follows from Lemma 4.11.

Next, we are going to prove that the number of fixed points of F, counted with weight given by the local degree, is equal to

$$w_w + b_b - a + b, \tag{4.19}$$

which, by (4.17) and the claim above, is equal to

$$\deg F + \deg(F|_{\mathscr{C}}) - (\deg(F|_{\mathscr{C}}) - 1) = 1 + \deg F.$$

Indeed, by Lemmas 4.6 and 4.7, each 1-tile that contributes in (4.19), i.e., each 1-tile in $\mathbf{X}^1_{ww} \cup \mathbf{X}^1_{bb} \cup B \cup A$, contains exactly one fixed point (not counted with weight) of F. On the other hand, each fixed point is contained in at least one of the 1-tiles in $\mathbf{X}^1_{ww} \cup \mathbf{X}^1_{bb} \cup B \cup A$. Let p be a fixed point of F, then one of the following happens:

1. If $p \notin \mathscr{C}$, then p is not contained in any 1-edges e since $F(e) \subseteq \mathscr{C}$. So $p \in \text{inte}(X)$ for some $X \in \mathbf{X}^1_{ww} \cup \mathbf{X}^1_{bb} \setminus (A \cup B)$, by Lemma 4.6. So each such p contributes 1 to (4.19).
2. If $p \in \mathscr{C}$ but $p \notin \text{crit}\, F$, then p is not a 1-vertex, so either p is contained in exactly two 1-tiles $X \in \mathbf{X}^1_{ww}$ and $X' \in \mathbf{X}^1_{bb}$, or p is contained in exactly two 1-tiles $X \in \mathbf{X}^1_{bw}$ and $X' \in \mathbf{X}^1_{wb}$. In either case, p contributes 1 to (4.19).
3. If $p \in \mathscr{C}$ and $p \in \text{crit}\, F$, then p is a 0-vertex, so the part that p contributes in (4.19) counts the number of black 1-tiles that contains p, which is exactly the weight $\deg_F(p)$ of p.

The proof is now complete. □

We can use some elementary number-theoretic argument to generalize Proposition 4.12 to all expanding Thurston maps, thus proving Theorem 4.1.

Proof (Proof of Theorem 4.1) We first observe that by Proposition 4.12, the theorem holds for f replaced by $F = f^n$ for each $n \geq N(f)$ where $N(f)$ is a constant as given in Corollary 2.18 depending only on f. Indeed, let $\mathscr{C}$ be an f^n-invariant Jordan curve containing $\text{post}\, f$ such that no n-tile in $\mathbf{D}^n(f, \mathscr{C})$ joins opposite sides of $\mathscr{C}$ as given in Corollary 2.18. So $\mathscr{C}$ is an F-invariant Jordan curve containing $\text{post}\, F$ such that no 1-tile in $\mathbf{D}^1(F, \mathscr{C})$ joins opposite sides of $\mathscr{C}$. Proposition 4.12 now applies.

Next, choose a prime $r \geq N(f)$. Note that the set of fixed points of f^r can be decomposed into periodic orbits under f of length r or 1, since r is a prime. Let p be a fixed point of f^r. By using the following formula derived from (2.2),

$$\deg_{f^r}(p) = \deg_f(p) \deg_f(f(p)) \deg_f(f^2(p)) \cdots \deg_f(f^{r-1}(p)), \tag{4.20}$$

we can conclude that

(1) if $p \notin \text{crit}\,(f^r)$, or equivalently, $\deg_{f^r}(p) = 1$, and

  (i) if p is in a periodic orbit of length r, then $p, f(p), \dots, f^{r-1}(p) \notin \text{crit}\, f$, or equivalently, the local degrees of f^r at these points are all 1;
  (ii) if p is in a periodic orbit of length 1, then $p \notin \text{crit}\, f$, or equivalently, $\deg_f(p) = 1$;

(2) if $p \in \text{crit}\,(f^r)$, and

  (i) if p is in a periodic orbit of length r, then all $p, f(p), \dots, f^{r-1}(p)$ are fixed points of f^r with the same weight $\deg_{f^r}(p) = \deg_{f^r}(f^k(p))$ for each $k \in \mathbb{N}$;
  (ii) if p is in a periodic orbit of length 1, then $p \in \text{crit}\, f$ and the weight of f^r at p is $\deg_{f^r}(p) = (\deg_f(p))^r$.

Note that a fixed point $p \in S^2$ of f^r is a fixed point of f if and only if p is in a periodic orbit of length 1 under f. So by first summing the weight of the fixed points of f^r in the same periodic orbit then summing over all such orbits and applying Fermat's Little Theorem, we can conclude that

$$\begin{aligned}
p_{1,f^r} &= \sum_{x \in P_{1,f^r}} \deg_{f^r}(x) \\
&= \sum_{(1)(i)} r + \sum_{(1)(ii)} 1 + \sum_{(2)(i)} r \deg_{f^r}(p) + \sum_{(2)(ii)} (\deg_f(p))^r \\
&\equiv \sum_{(1)(ii)} 1 + \sum_{(2)(ii)} \deg_f(p) \\
&= p_{1,f} \pmod{r},
\end{aligned}$$

where on the second line, the first sum ranges over all periodic orbits in Case (1)(i), the second sum ranges over all periodic orbits in Case (1)(ii), the third sum ranges over all periodic orbits of the form $\{p, f(p), \dots, f^{r-1}(p)\}$ in Case (2)(i), the last sum ranges over all periodic orbits of the form $\{p\}$ in Case (2)(ii). Thus by (2.3) and Fermat's Little Theorem again, we have

$$0 = \deg(f^r) + 1 - p_{1,f^r} \equiv (\deg f)^r + 1 - p_{1,f} \equiv 1 + \deg f - p_{1,f} \pmod{r}. \quad (4.21)$$

By choosing the prime r larger than $|1 + \deg f - p_{1,f}|$, we can conclude that $p_{1,f} = 1 + \deg f$. □

In particular, we have the following corollary in which the weight for all points are trivial.

Corollary 4.13 *If f is an expanding Thurston map with no critical fixed points, then there are exactly $1 + \deg f$ distinct fixed points of f. Moreover, if f is an expanding Thurston map with no periodic critical points, then there are exactly $1 + (\deg f)^n$ distinct fixed points of f^n, for each $n \in \mathbb{N}$.*

Proof The first statement follows immediately from Theorem 4.1.

To prove the second statement, we first recall that if f is an expanding Thurston map, so is f^n for each $n \in \mathbb{N}$. Next we note that for each fixed point $p \in S^2$ of f^n, $n \in \mathbb{N}$, we have $\deg_{f^n}(p) = 1$. For otherwise, suppose $\deg_{f^n}(p) > 1$ for some $n \in \mathbb{N}$, then

$$1 < \deg_{f^n}(p) = \deg_f(p) \deg_f(f(p)) \deg_f(f^2(p)) \cdots \deg_f(f^{n-1}(p)).$$

Thus at least one of the points $p, f(p), f^2(p), \dots, f^{n-1}(p)$ is a periodic critical point of f, a contradiction. The second statement now follows. □

We recall the definition of s_n^m in (4.15) and (4.16).

Corollary 4.14 *Let f be an expanding Thurston map. For each $m \in \mathbb{N}_0$ and $n \in \mathbb{N}$ with $m < n$, we have*

$$s_n^m = (\deg f)^n + (\deg f)^m. \tag{4.22}$$

Proof For all $m \in \mathbb{N}_0$ and $n \in \mathbb{N}$ with $m < n$, we have

$$\begin{aligned} s_n^m &= \sum_{x \in S_n^m} \deg_{f^n}(x) = \sum_{y = f^{n-m}(y)} \sum_{x \in f^{-m}(y)} \deg_{f^n}(x) \\ &= \sum_{y = f^{n-m}(y)} \deg_{f^{n-m}}(y) \sum_{x \in f^{-m}(y)} \deg_{f^m}(x) \\ &= \left((\deg f)^{n-m} + 1\right)(\deg f)^m. \end{aligned}$$

The last equality follows from (2.1), (2.3), and Theorem 4.1. □

Finally, for expanding Thurston maps with no periodic critical points, we derive a formula for $p_{n,f}$, $n \in \mathbb{N}$, from Theorem 4.1 and the Möbius inversion formula (see for example, [Bak85, Sect. 2.4]).

Definition 4.15 The *Möbius function*, $\mu(u)$, is defined by

$$\mu(n) = \begin{cases} 1 & \text{if } n = 1; \\ (-1)^r & \text{if } n = p_1 p_2 \dots p_r, \text{ and } p_1, \dots, p_r \text{ are distinct primes;} \\ 0 & \text{otherwise.} \end{cases}$$

Corollary 4.16 *Let f be an expanding Thurston map without any periodic critical points. Then for each $n \in \mathbb{N}$, we have*

$$p_{n,f} = \sum_{d|n} \mu(d) p_{1,f^{n/d}} = \begin{cases} \sum_{d|n} \mu(d)(\deg f)^{n/d} & \text{if } n > 1; \\ 1 + \deg f & \text{if } n = 1. \end{cases}$$

Proof The first equality follows from the Möbius inversion formula and the equation $p_{1,f^n} = \sum_{d|n} p_{d,f}$, for $n \in \mathbb{N}$. The second equality follows from Theorem 4.1 and the following fact (see for example, [Bak85, Sect. 2.4]):

$$\sum_{d|n} \mu(d) = \begin{cases} 1 & \text{if } n = 1, \\ 0 & \text{if } n > 1. \end{cases}$$

□

4.2 Equidistribution

In this section, we derive various equidistribution results as stated in Theorems 4.2, 4.3, and Corollary 4.4. We prove these results by first establishing a general statement in Theorem 4.23 on the convergence of the distributions of the white n-tiles in the tile decompositions discussed in Sect. 2.2, in the weak* topology, to the unique measure of maximal entropy of an expanding Thurston map. More precisely, we show in Theorem 4.23 that the distributions of the points, each of which is located "near" its corresponding white n-tile where the correspondence is a bijection, converges in the weak* topology to μ_f as $n \longrightarrow +\infty$. Then Theorem 4.2 follows from Theorem 4.23. Theorem 4.3 finally follows after we prove a technical bound in Lemma 4.28 generalizing a corresponding lemma from [BM17]. As a special case, we obtain Corollary 4.4. Theorem 4.1 is used in several places in this section.

Let f be an expanding Thurston map. Then

$$h_{\mathrm{top}}(f) = \log(\deg f), \tag{4.23}$$

and there exists a unique measure of maximal entropy μ_f for f (see [BM17, Theorem 17.11] and [HP09, Sects. 3.4 and 3.5]). Moreover, for each $n \in \mathbb{N}$, the unique measure of maximal entropy μ_{f^n} of the expanding Thurston map f^n is equal to μ_f (see [BM17, Proposition 17.10 and Theorem 17.11]).

We recall that in a compact metric space (X, d), a sequence of finite Borel measures μ_n converges in the weak* topology to a finite Borel measure μ, or $\mu_n \xrightarrow{w^*} \mu$, as $n \longrightarrow +\infty$ if and only if $\lim\limits_{n \to +\infty} \int u \,\mathrm{d}\mu_n = \int u \,\mathrm{d}\mu$ for each $u \in C(X)$.

We need the following lemmas for weak* convergence.

Lemma 4.17 *Let X and $\widetilde{X}$ be two compact metric spaces and $\phi\colon X \to \widetilde{X}$ a continuous map. Let μ and μ_i, for $i \in \mathbb{N}$, be finite Borel measures on X. If*

$$\mu_i \xrightarrow{w^*} \mu \text{ as } i \longrightarrow +\infty,$$

then $\phi_*(\mu)$ and $\phi_*(\mu_i)$, $i \in \mathbb{N}$, are finite Borel measures on $\widetilde{X}$, and

$$\phi_*(\mu_i) \xrightarrow{w^*} \phi_*(\mu) \text{ as } i \longrightarrow +\infty.$$

Recall for a continuous map $\phi\colon X \to \widetilde{X}$ between two metric spaces and a Borel measure ν on X, the *push-forward* $\phi_*(\nu)$ of ν by ϕ is defined to be the unique Borel measure that satisfies $(\phi_*(\nu))(B) = \nu\left(\phi^{-1}(B)\right)$ for each Borel set $B \subseteq \widetilde{X}$.

Proof By the Riesz representation theorem (see for example, [Fo99, Chap. 7]), the lemma follows if we observe that for each $h \in C(\widetilde{X})$, we have

$$\int_{\widetilde{X}} h \, \mathrm{d}\phi_*(\mu_i) = \int_X (h \circ \phi) \, \mathrm{d}\mu_i \overset{i \longrightarrow +\infty}{\longrightarrow} \int_X (h \circ \phi) \, \mathrm{d}\mu = \int_{\widetilde{X}} h \, \mathrm{d}\phi_*(\mu).$$

□

Lemma 4.18 *Let (X, d) be a compact metric space, and I be a finite set. Suppose that μ and $\mu_{i,n}$, for $i \in I$ and $n \in \mathbb{N}$, are finite Borel measures on X, and $w_{i,n} \in [0, +\infty)$, for $i \in I$ and $n \in \mathbb{N}$ such that*

1. $\mu_{i,n} \overset{w^*}{\longrightarrow} \mu$, *as* $n \longrightarrow +\infty$, *for each* $i \in I$,
2. $\lim\limits_{n \to +\infty} \sum\limits_{i \in I} w_{i,n} = r$ *for some* $r \in \mathbb{R}$.

Then $\sum\limits_{i \in I} w_{i,n}\mu_{i,n} \overset{w^*}{\longrightarrow} r\mu$ *as* $n \longrightarrow +\infty$.

Proof For each $u \in C(X)$ and each $n \in \mathbb{N}$,

$$\left| \int u \, \mathrm{d}\Big(\sum_{i \in I} w_{i,n}\mu_{i,n} \Big) - r \int u \, \mathrm{d}\mu \right| \leq \sum_{i \in I} w_{i,n} \left| \int u \, \mathrm{d}\mu_{i,n} - \int u \, \mathrm{d}\mu \right| + \left| r - \sum_{i \in I} w_{i,n} \right| \int |u| \, \mathrm{d}\mu.$$

Since $\mu_{i,n} \overset{w^*}{\longrightarrow} \mu$, as $n \longrightarrow +\infty$, for each $i \in I$, and $\lim\limits_{n \to +\infty} \sum\limits_{i \in I} w_{i,n} = r$, we can conclude that the right-hand side of the inequality above tends to 0 as $n \longrightarrow +\infty$. □

We record the following well-known lemma, sometimes known as the Portmanteau Theorem, and refer the reader to [Bi99, Theorem 2.1] for the proof.

Lemma 4.19 *Let (X, d) be a compact metric space, and μ and μ_i, for $i \in \mathbb{N}$, be Borel probability measures on X. Then the following are equivalent:*

1. $\mu_i \overset{w^*}{\longrightarrow} \mu$ *as* $i \longrightarrow +\infty$;
2. $\limsup\limits_{i \to +\infty} \mu_i(F) \leq \mu(F)$ *for each closed set* $F \subseteq X$;
3. $\liminf\limits_{i \to +\infty} \mu_i(G) \geq \mu(G)$ *for each open set* $G \subseteq X$;
4. $\lim\limits_{i \to +\infty} \mu_i(B) = \mu(B)$ *for each Borel set* $B \subseteq X$ *with* $\mu(\partial B) = 0$.

Lemma 4.20 *Let (X, d) be a compact metric space. Suppose that $A_i \subseteq X$, for $i \in \mathbb{N}$, are finite subsets of X with maps $\phi_i \colon A_i \to X$ such that*

$$\lim_{i \to +\infty} \max\{d(x, \phi_i(x)) \,|\, x \in A_i\} = 0.$$

Let $m_i \colon A_i \to \mathbb{R}$, for $i \in \mathbb{N}$, be functions that satisfy

$$\sup_{i\in\mathbb{N}} \|m_i\|_1 = \sup_{i\in\mathbb{N}} \sum_{x\in A_i} |m_i(x)| < +\infty.$$

Define for each $i \in \mathbb{N}$,

$$\mu_i = \sum_{x\in A_i} m_i(x)\delta_x, \quad \widetilde{\mu}_i = \sum_{x\in A_i} m_i(x)\delta_{\phi_i(x)}.$$

If

$$\mu_i \xrightarrow{w^*} \mu \text{ as } i \longrightarrow +\infty,$$

for some finite Borel measure μ *on* X, *then*

$$\widetilde{\mu}_i \xrightarrow{w^*} \mu \text{ as } i \longrightarrow +\infty.$$

Proof It suffices to prove that for each continuous function $g \in C(X)$,

$$\int g \,\mathrm{d}\mu_i - \int g \,\mathrm{d}\widetilde{\mu}_i \longrightarrow 0 \text{ as } i \longrightarrow +\infty.$$

Indeed, g is uniformly continuous, so for each $\varepsilon > 0$, there exists $N \in \mathbb{N}$ such that for each $n > N$ and for each $x \in A_n$, we have $|g(x) - g(\phi_n(x))| < \varepsilon$. Thus

$$\left| \int g \,\mathrm{d}\mu_n - \int g \,\mathrm{d}\widetilde{\mu}_n \right| \leq \sum_{x\in A_n} |g(x) - g(\phi_n(x))| \, |m_n(x)| \leq \varepsilon \sup_{n\in\mathbb{N}} \|m_n\|_1 .$$

□

The following lemma is a reformulation of Lemma 17.6 in [BM17]. We will later generalize it in Lemma 4.28.

Lemma 4.21 (M. Bonk and D. Meyer) *Let* f *be an expanding Thurston map, and* $\mathscr{C} \subseteq S^2$ *be an* f^N*-invariant Jordan curve containing* post f *for some* $N \in \mathbb{N}$. *Then there exists a constant* $L_0 \in [1, \deg f)$ *with the following property:*

For each $m \in \mathbb{N}_0$ *with* $m \equiv 0 \pmod N$, *there exists a constant* $D_0 > 0$ *such that for each* $k \in \mathbb{N}_0$ *with* $k \equiv 0 \pmod N$ *and each* m*-edge* e, *there exists a collection* M_0 *of* $(m+k)$*-tiles with* card $M_0 \leq D_0 L_0^k$ *and* $e \subseteq \operatorname{int}\left(\bigcup_{X\in M_0} X \right)$.

Let F be an expanding Thurston map with an F-invariant Jordan curve $\mathscr{C} \subseteq S^2$ containing post F. As before, we let $w_w = \operatorname{card} \mathbf{X}^1_{ww}$ denote the number of white 1-tiles contained in the white 0-tile, $b_w = \operatorname{card} \mathbf{X}^1_{bw}$ the number of black 1-tiles contained in the white 0-tile, $w_b = \operatorname{card} \mathbf{X}^1_{wb}$ the number of white 1-tiles contained in the black 0-tile, and $b_b = \operatorname{card} \mathbf{X}^1_{bb}$ the number of black 1-tiles contained in the black 0-tile. We define

$$w = \frac{b_w}{b_w + w_b}, \quad b = \frac{w_b}{b_w + w_b}. \tag{4.24}$$

Note that (see the discussion in [BM17] proceeding Lemma 17.3 in Chap. 17) $b_w, w_b, w, b > 0$, $w + b = 1$, and

$$|w_w - b_w| < \deg F. \tag{4.25}$$

M. Bonk and D. Meyer gave the following characterization of the unique measure of maximal entropy of F (see [BM17, Proposition 17.10 and Theorem 17.11]):

Theorem 4.22 (M. Bonk and D. Meyer) *Let F be an expanding Thurston map with an F-invariant Jordan curve $\mathcal{C} \subseteq S^2$. Then there is a unique measure of maximal entropy μ_F of F, which is characterized among all Borel probability measures by the following property:*

For each $n \in \mathbb{N}_0$ and each n-tile $X^n \in \mathbf{X}^n(F, \mathcal{C})$,

$$\mu_F(X^n) = \begin{cases} w(\deg F)^{-n} & \text{if } X^n \in \mathbf{X}^n_w(F, \mathcal{C}), \\ b(\deg F)^{-n} & \text{if } X^n \in \mathbf{X}^n_b(F, \mathcal{C}). \end{cases} \tag{4.26}$$

We now state our first characterization of the measure of maximal entropy μ_f of an expanding Thurston map f.

Theorem 4.23 *Let f be an expanding Thurston map with its measure of maximal entropy μ_f. Let $\mathcal{C} \subseteq S^2$ be an f^n-invariant Jordan curve containing* post f *for some $n \in \mathbb{N}$. Fix a visual metric d for f. Consider any sequence of non-negative numbers $\{\alpha_i\}_{i \in \mathbb{N}_0}$ with $\lim_{i \to +\infty} \alpha_i = 0$, and any sequence of maps $\{\beta_i\}_{i \in \mathbb{N}_0}$ with β_i sending each white i-tile $X^i \in \mathbf{X}^i_w(f, \mathcal{C})$ to a point $\beta_i(X^i) \in N^{\alpha_i}_d(X^i)$. Let*

$$\mu_i = \frac{1}{(\deg f)^i} \sum_{X^i \in \mathbf{X}^i_w(f, \mathcal{C})} \delta_{\beta_i(X^i)}, \quad i \in \mathbb{N}_0.$$

Then

$$\mu_i \xrightarrow{w^*} \mu_f \text{ as } i \longrightarrow +\infty.$$

Recall that $N^{\alpha_i}_d(X^i)$ denotes the open α_i-neighborhood of X^i in (S^2, d). This theorem says that a sequence of probability measures $\{\mu_i\}_{i \in \mathbb{N}}$, with μ_i assigning the same weight to a point near each white i-tile, converges in the weak* topology to the measure of maximal entropy. In some sense, it asserts the equidistribution of the white i-tiles with respect to the measure of maximal entropy.

We first prove a weaker version of the above theorem.

Proposition 4.24 *Let F be an expanding Thurston map with its measure of maximal entropy μ_F and an F-invariant Jordan curve $\mathscr{C} \subseteq S^2$ containing* post F. *Consider any sequence of maps $\{\beta_i\}_{i \in \mathbb{N}_0}$ with β_i sending each white i-tile $X^i \in \mathbf{X}^i_w(F, \mathscr{C})$ to a point $\beta_i(X^i) \in \operatorname{inte}(X^i)$ for each $i \in \mathbb{N}_0$. Let*

$$\mu_i = \frac{1}{(\deg F)^i} \sum_{X^i \in \mathbf{X}^i_w(F,\mathscr{C})} \delta_{\beta_i(X^i)}, \quad i \in \mathbb{N}_0.$$

Then

$$\mu_i \xrightarrow{w^*} \mu_F \text{ as } i \longrightarrow +\infty.$$

Proof Note that card $\mathbf{X}^i_w = (\deg F)^i$, so μ_i is a probability measure for each $i \in \mathbb{N}_0$. Thus by Alaoglu's theorem, it suffices to prove that for each Borel measure μ which is a subsequential limit of $\{\mu_i\}_{i \in \mathbb{N}_0}$ in the weak* topology, we have $\mu = \mu_F$.

Let $\{i_n\}_{n \in \mathbb{N}} \subseteq \mathbb{N}$ be an arbitrary strictly increasing sequence such that

$$\mu_{i_n} \xrightarrow{w^*} \mu \text{ as } n \longrightarrow +\infty,$$

for some Borel measure μ. Clearly μ is also a probability measure.

Recall the definitions of $w, b \in (0,1)$ and w_w, b_w, w_b, b_b (see (4.24)). For each $m, i \in \mathbb{N}_0$ with $0 \leq m \leq i$, each white m-tile $X^m_w \in \mathbf{X}^m_w$, and each black m-tile $X^m_b \in \mathbf{X}^m_b$, by the formulas in Lemma 17.3 in [BM17], we have

$$\begin{aligned} \mu_i(X^m_w) &= \frac{1}{(\deg F)^i} \operatorname{card}\{X^i \in \mathbf{X}^i_w \mid X^i \subseteq X^m_w\} \\ &= \frac{1}{(\deg F)^i} \left(w(\deg F)^{i-m} + b(w_w - b_w)^{i-m} \right), \end{aligned} \tag{4.27}$$

and similarly,

$$\mu_i(X^m_b) = \frac{1}{(\deg F)^i} \left(w(\deg F)^{i-m} - b(w_w - b_w)^{i-m} \right). \tag{4.28}$$

We claim that for each m-tile $X^m \in \mathbf{X}^m$ with $m \in \mathbb{N}_0$, we have $\mu(\partial X^m) = 0$.

To establish the claim, by Proposition 2.6(vi), it suffices to prove that $\mu(e) = 0$ for each m-edge e with $m \in \mathbb{N}_0$. Applying Lemma 4.21 in the case $f = F$ and $n = 1$, we get that there exist constants $1 < L_0 < \deg F$ and $D_0 > 0$ such that for each $k \in \mathbb{N}_0$, there is a collection M^k_0 of $(m+k)$-tiles with card $M^k_0 \leq D_0 L^k_0$ such that e is contained in the interior of the set $\bigcup_{X \in M^k_0} X$. So by (4.25), (4.27), (4.28), and Lemma 4.19, we get

$$\mu(e) \le \mu\Big(\operatorname{int}\Big(\bigcup_{X \in M_0^k} X\Big)\Big) \le \limsup_{l \to +\infty} \mu_{m+k+l}\Big(\operatorname{int}\Big(\bigcup_{X \in M_0^k} X\Big)\Big)$$

$$\le \limsup_{l \to +\infty} \sum_{X \in M_0^k} \mu_{m+k+l}(X) \le \sum_{X \in M_0^k} \limsup_{l \to +\infty} \mu_{m+k+l}(X) \le D_0 L_0^k \frac{w+b}{(\deg F)^{m+k}}.$$

By letting $k \longrightarrow +\infty$, we get $\mu(e) = 0$, proving the claim.

Thus by (4.25), (4.27), (4.28), the claim, and Lemma 4.19, we can conclude that for each $m \in \mathbb{N}_0$, and each white m-tile $X_w^m \in \mathbf{X}_w^m$, each black m-tile $X_b^m \in \mathbf{X}_b^m$, we have that

$$\mu(X_w^m) = \lim_{n \to +\infty} \mu_{i_n}(X_w^m) = w(\deg F)^{-m},$$

$$\mu(X_b^m) = \lim_{n \to +\infty} \mu_{i_n}(X_b^m) = b(\deg F)^{-m}.$$

By Theorem 4.22, therefore, the measure μ is equal to the unique measure of maximal entropy μ_F of F. □

As a consequence of the above proposition, we have

Corollary 4.25 *Let f be an expanding Thurston map with its measure of maximal entropy μ_f. Let $\mathcal{C} \subseteq S^2$ be an f^n-invariant Jordan curve containing* post f *for some $n \in \mathbb{N}$. Fix an arbitrary $p \in \operatorname{inte}(X_w^0)$ where X_w^0 is the white 0-tile for $(f, \mathcal{C})$. Define, for $i \in \mathbb{N}$,*

$$\nu_i = \frac{1}{(\deg f)^i} \sum_{q \in f^{-i}(p)} \delta_q.$$

Then

$$\nu_i \xrightarrow{w^*} \mu_f \text{ as } i \longrightarrow +\infty.$$

Proof First observe that since p is contained in the interior of the white 0-tile, each $q \in f^{-n}(p)$ is contained in the interior of one of the white n-tiles, and each white n-tile contains exactly one q with $f^n(q) = p$. So by Proposition 4.24,

$$\nu_{ni} \xrightarrow{w^*} \mu_{f^n} \text{ as } i \longrightarrow +\infty, \tag{4.29}$$

where μ_{f^n} is the unique measure of maximal entropy of f^n, which is equal to μ_f (see [BM17, Proposition 17.10 and Theorem 17.11]).

Then note that for $k > 1$,

$$f_* \nu_k = \frac{1}{(\deg f)^k} \sum_{q \in f^{-k}(p)} \delta_{f(q)} = \frac{1}{(\deg f)^{k-1}} \sum_{q \in f^{-k+1}(p)} \delta_q = \nu_{k-1}. \tag{4.30}$$

The second equality above follows from the fact that the number of preimages of each point in $f^{-k+1}(p)$ is exactly $\deg f$.

So by (4.29), (4.30), Lemma 4.17, and the fact that μ_f is invariant under push-forward of f from Theorem 17.11 in [BM17], for each $k \in \{0, 1, \dots, n-1\}$, we get

$$\nu_{ni-k} = (f_*)^k \nu_{ni} \xrightarrow{w^*} (f_*)^k \mu_f = \mu_f \text{ as } i \longrightarrow +\infty.$$

Therefore

$$\nu_i \xrightarrow{w^*} \mu_f \text{ as } i \longrightarrow +\infty.$$

□

Proof (Proof of Theorem 4.23) Fix an arbitrary $p \in \operatorname{inte}(X_w^0)$ in the interior of the while 0-tile X_w^0 for the cell decomposition induced by $(f, \mathscr{C})$.

As in the proof of Corollary 4.25, for each $i \in \mathbb{N}_0$, there is a bijective correspondence between points in $f^{-i}(p)$ and the set of white i-tiles, namely, each $q \in f^{-i}(p)$ corresponds to the unique white i-tile, denoted by X_q, containing q. Then we define functions $\phi_i : f^{-i}(p) \to S^2$ by setting $\phi_i(q) = \beta_i(X_q)$.

Let $\Lambda > 1$ be the expansion factor of our fixed visual metric d. There exists $C \geq 1$ such that for each $n \in \mathbb{N}_0$ and each n-tile $X^n \in \mathbf{X}^n$, $\operatorname{diam}_d(X^n) \leq C\Lambda^{-n}$ (see Lemma 2.13). So for each $i \in N_0$ and each $q \in f^{-i}(p)$, we have

$$d(q, \phi_i(q)) \leq d(\phi_i(q), X_q) + \operatorname{diam}_d(X_q) \leq \alpha_i + C\Lambda^{-i}.$$

Thus $\lim\limits_{i \to +\infty} \max\{d(x, \phi_i(x)) \mid x \in f^{-i}(p)\} = 0$.

For $i \in \mathbb{N}_0$, define

$$\widetilde{\mu}_i = \frac{1}{(\deg f)^i} \sum_{q \in f^{-i}(p)} \delta_q.$$

Note that for $i \in \mathbb{N}_0$,

$$\mu_i = \frac{1}{(\deg f)^i} \sum_{X^i \in \mathbf{X}_w^i(f, \mathscr{C})} \delta_{\beta_i(X^i)} = \frac{1}{(\deg f)^i} \sum_{q \in f^{-i}(p)} \delta_{\phi_i(q)}.$$

Then by Corollary 4.25,

$$\widetilde{\mu}_i \xrightarrow{w^*} \mu_f \text{ as } i \longrightarrow +\infty.$$

Therefore, by Lemma 4.20 with $A_i = f^{-i}(p)$ and $m_i(x) = \frac{1}{(\deg f)^i}$, $i \in \mathbb{N}_0$, we can conclude that

$$\mu_i \xrightarrow{w^*} \mu_f \text{ as } i \longrightarrow +\infty.$$

□

Remark 4.26 We can replace "white" by "black", $\mathbf{X}_w^i$ by $\mathbf{X}_b^i$, and X_w^0 by X_b^0 in the statements of Theorem 4.23, Proposition 4.24, and Corollary 4.25. The proofs are essentially the same.

We are now ready to prove the equidistribution of preimages of an arbitrary point with respect to the measure of maximal entropy μ_f.

Proof (Proof of Theorem 4.2) By Theorem 15.1 in [BM17] or Corollary 2.18, we can fix an f^n-invariant Jordan curve $\mathscr{C} \subseteq S^2$ containing post f for some $n \in \mathbb{N}$. We consider the cell decompositions induced by $(f, \mathscr{C})$.

We first prove (4.2).

We assume that p is contained in the (closed) white 0-tile. The proof for the case when p is contained in the black 0-tile is exactly the same except that we need to use a version of Theorem 4.23 for black tiles instead of using Theorem 4.23 literally, see Remark 4.26.

Observe that for each $i \in \mathbb{N}_0$ and each $q \in f^{-i}(p)$, the number of white i-tiles that contains q is exactly $\deg_{f^i}(q)$. On the other hand, each white i-tile contains exactly one point q with $f^i(q) = p$. So we can define $\beta_i \colon \mathbf{X}_w^i \to S^2$ by mapping a white i-tile to the point q in it that satisfies $f^i(q) = p$. Define $\alpha_i \equiv 0$. Theorem 4.23 applies, and thus (4.2) is true.

Next, we prove (4.3). The proof breaks into three cases.

Case 1. Assume that $p \notin \operatorname{post} f$. Then $\deg_f(x) = 1$ for all $x \in \bigcup_{n=1}^{+\infty} f^{-n}(p)$. So $\widetilde{\nu}_i = \nu_i$ for each $i \in \mathbb{N}$. Then (4.3) follows from (4.2) in this case.

Case 2. Assume that $p \in \operatorname{post} f$ and p is not periodic. Then there exists $N \in \mathbb{N}$ such that $f^{-N}(p) \cap \operatorname{post} f = \emptyset$. For otherwise, there exists a point $z \in \operatorname{post} f$ which belongs to $f^{-c}(p)$ for infinitely many distinct $c \in \mathbb{N}$. In particular, there exist two integers $a > b > 0$ such that $z \in f^{-a}(p) \cap f^{-b}(p)$. Then $f^{a-b}(p) = p$, a contradiction. So $\deg_f(q) = 1$ for each $q \in \bigcup_{x \in f^{-N}(p)} \bigcup_{i=1}^{+\infty} f^{-i}(x)$. Note that for each $x \notin \operatorname{post} f$ and each $i \in \mathbb{N}$, the number of preimages of x under f^i is exactly $(\deg f)^i$. Then for each $i \in \mathbb{N}$, $Z_{i+N} = Z_N (\deg f)^i$, and

$$\widetilde{\nu}_{i+N} = \frac{1}{Z_{i+N}} \sum_{q \in f^{-(i+N)}(p)} \delta_q = \frac{1}{Z_N} \sum_{x \in f^{-N}(p)} \left(\frac{1}{(\deg f)^i} \sum_{q \in f^{-i}(x)} \delta_q \right).$$

For each $x \in f^{-N}(p)$, by Case 1,

$$\frac{1}{(\deg f)^i} \sum_{q \in f^{-i}(x)} \delta_q \xrightarrow{w^*} \mu_f \text{ as } i \longrightarrow +\infty.$$

Thus each term in the sequence $\{\widetilde{\nu}_{i+N}\}_{i \in \mathbb{N}}$ is a convex combination of the corresponding terms in sequences of measures, each of which converges to μ_f in the weak* topology. Hence by Lemma 4.18, the sequence $\{\widetilde{\nu}_{i+N}\}_{i\in\mathbb{N}}$ also converges to μ_f in the weak* topology in this case.

Case 3. Assume that $p \in \operatorname{post} f$ and p is periodic with period $k \in \mathbb{N}$. Let $l = \operatorname{card}(\operatorname{post} f)$. We first note that for each $m, N \in \mathbb{N}$, the inequality

$$Z_{m+N} \geq (Z_m - l)(\deg f)^N,$$

and equivalently,

$$\frac{Z_m}{Z_{m+N}} \leq \frac{1}{(\deg f)^N} + \frac{l}{Z_{m+N}}$$

hold, since there are at most l points in $Z_m \cap \operatorname{post} f$. So by Lemma 2.12, for each $\varepsilon > 0$ and each N large enough such that $1/(\deg f)^N < \varepsilon/2$ and $l/Z_{m+N} < \varepsilon/2$, we get $Z_m/Z_{m+N} < \varepsilon$ for each $m \in \mathbb{N}$. We fix $j \in \mathbb{N}$ large enough such that $Z_{m-jk}/Z_m < \varepsilon$ for each $m > jk$. Observe that for each $m > jk$,

$$\begin{aligned}
\widetilde{\nu}_m &= \frac{1}{Z_m}\sum_{q\in f^{-m}(p)} \delta_q = \frac{1}{Z_m}\Bigg(\sum_{q\in f^{-(m-jk)}(p)} \delta_q + \sum_{x\in f^{-jk}(p)\setminus\{p\}} \sum_{q\in f^{-(m-jk)}(x)} \delta_q\Bigg) \\
&= \frac{Z_{m-jk}}{Z_m}\Bigg(\frac{1}{Z_{m-jk}}\sum_{q\in f^{-(m-jk)}(p)} \delta_q\Bigg) + \frac{1}{Z_m}\sum_{x\in f^{-jk}(p)\setminus\{p\}} \qquad (4.31)\\
&\quad \operatorname{card}\big(f^{-(m-jk)}(x)\big)\Bigg(\frac{1}{\operatorname{card}\big(f^{-(m-jk)}(x)\big)}\sum_{q\in f^{-(m-jk)}(x)} \delta_q\Bigg).
\end{aligned}$$

Note that no point $x \in f^{-jk}(p)\setminus\{p\}$ is periodic. Indeed, if $x \in f^{-jk}(p)\setminus\{p\}$ were periodic, then $x \in \bigcup_{i=0}^{k-1} f^i(p)$, and so x would have period k as well. Thus $x = f^{jk}(x) = p$, a contradiction. Hence by Case 1 and Case 2, for each $x \in f^{-jk}(p)\setminus\{p\}$,

$$\frac{1}{\operatorname{card}\big(f^{-(m-jk)}(x)\big)}\sum_{q\in f^{-(m-jk)}(x)} \delta_q \xrightarrow{w^*} \mu_f \text{ as } m \longrightarrow +\infty.$$

Let $\mu \in \mathscr{P}(S^2)$ be an arbitrary subsequential limit of $\{\widetilde{\nu}_m\}_{m\in\mathbb{N}}$ in the weak* topology. For each strictly increasing sequence $\{m_i\}_{i\in\mathbb{N}}$ in $\mathbb{N}$ that satisfies

$$\widetilde{\nu}_{m_i} \xrightarrow{w^*} \mu \text{ as } i \longrightarrow +\infty,$$

we can assume, due to Alaoglu's Theorem, by choosing a subsequence if necessary, that

$$\frac{Z_{m_i-jk}}{Z_{m_i}}\Bigg(\frac{1}{Z_{m_i-jk}}\sum_{q\in f^{-(m_i-jk)}(p)} \delta_q\Bigg) \xrightarrow{w^*} \eta \text{ as } i \longrightarrow +\infty,$$

for some Borel measure η with total variation $\|\eta\| \leq \varepsilon$. Observe that for each $i \in \mathbb{N}$,

$$\frac{1}{Z_{m_i}} \sum_{x \in f^{-jk}(p) \setminus \{p\}} \operatorname{card}\left(f^{-(m_i - jk)}(x)\right) = 1 - \frac{Z_{m_i - jk}}{Z_{m_i}},$$

since $p \in f^{-jk}(p)$ and $\operatorname{card}\left(f^{-(m_i-jk)}(p)\right) = Z_{m_i-jk}$. By choosing a subsequence of $\{m_i\}_{i \in \mathbb{N}}$ if necessary, we can assume that there exists $r \in [0, \varepsilon]$ such that

$$\lim_{i \to +\infty} \frac{Z_{m_i - jk}}{Z_{m_i}} = r.$$

So by taking the limits of both sides of (4.31) in the weak* topology along the subsequence $\{m_i\}_{i \in \mathbb{N}}$, we get from Lemma 4.18 that $\mu = \eta + (1 - r)\mu_f$. Thus

$$\left\|\mu - \mu_f\right\| \leq \|\eta\| + r \left\|\mu_f\right\| \leq 2\varepsilon.$$

Since ε is arbitrary, we can conclude that $\mu = \mu_f$. We have proved in this case that each subsequential limit of $\{\widetilde{\nu}_m\}_{m \in \mathbb{N}}$ in the weak* topology is equal to μ_f. Therefore (4.3) is true in this case. □

In order to prove Theorem 4.3, we will need Lemma 4.28 which is a generalization of Lemma 4.21.

Lemma 4.27 *Let f be an expanding Thurston map and $d = \deg f$. Then there exist constants $C > 0$ and $\alpha \in (0, 1]$ such that for each nonempty finite subset M of S^2 and each $n \in \mathbb{N}$, we have*

$$\frac{1}{d^n} \sum_{x \in M} \deg_{f^n}(x) \leq C \max \left\{ \left(\frac{\operatorname{card} M}{d^n}\right)^{\alpha}, \frac{\operatorname{card} M}{d^n} \right\}. \tag{4.32}$$

Note that when $\operatorname{card} M \leq d^n$, the right-hand side of (4.32) becomes $C\left(\frac{\operatorname{card} M}{d^n}\right)^{\alpha}$.

Proof Let $m = \operatorname{card} M$. Set

$$D = \prod_{x \in \operatorname{crit} f} \deg_f(x).$$

In order to establish the lemma, we consider the following three cases.

Case 1. Suppose that f has no periodic critical points. Then since for each $x \in S^2$ and each $n \in \mathbb{N}$,

$$\deg_{f^n}(x) = \deg_f(x) \deg_f(f(x)) \cdots \deg_f(f^{n-1}(x)), \tag{4.33}$$

it is clear that $\deg_{f^n}(x) \leq D$. So

$$\frac{1}{d^n}\sum_{x\in M}\deg_{f^n}(x)\leq D\frac{m}{d^n}.$$

Thus in this case, $C = D$ and $\alpha = 1$.

Case 2. Suppose that f has periodic critical points, but all periodic critical points are fixed points of f.

Let $T_0 = \{x \in \operatorname{crit} f \mid f(x) = x\}$ be the set of periodic critical points of f. Then define recursively for each $i \in \mathbb{N}$,

$$T_i = f^{-1}(T_{i-1})\setminus\bigcup_{j=0}^{i-1}T_j.$$

Define $T_{-1} = S^2\setminus\bigcup_{j=0}^{+\infty}T_j$, and $\widetilde{T}_i = S^2\setminus\bigcup_{j=0}^{i}T_j$ for each $i \in \mathbb{N}_0$. Set $t_0 = \operatorname{card} T_0$. Since $T_0 \subseteq \operatorname{post} f$, we have $1 \leq t_0 < +\infty$. Then for each $i \in \mathbb{N}$, we have

$$\operatorname{card} T_i \leq d^i t_0.$$

We note that if $\deg_f(x) = d$ for some $x \in T_0$, then $f^{-i}(x) = \{x\}$ for each $i \in \mathbb{N}$, contradicting Lemma 2.12. So $\deg_f(x) \leq d-1$ for each $x \in T_0$. Thus for each $x \in T_0$ and each $m \in \mathbb{N}$, we have

$$\deg_{f^m}(x) \leq (d-1)^m.$$

Moreover, for each $i, m \in \mathbb{N}$ with $i < m$ and each $x \in T_i$, we get

$$\deg_{f^m}(x) = \deg_f(x)\deg_f(f(x))\cdots\deg_f(f^{i-1}(x))\deg_{f^{m-i}}(f^i(x)) \leq D(d-1)^{m-i}.$$

Similarly, for each $i, m \in \mathbb{N}$ with $i \geq m$ and each $x \in \widetilde{T}_i$, we have

$$\deg_{f^m}(x) \leq D.$$

Thus for each $n \in \mathbb{N}$,

$$\begin{aligned}\frac{1}{d^n}\sum_{x\in M}\deg_{f^n}(x) &= \frac{1}{d^n}\sum_{j=-1}^{+\infty}\sum_{x\in M\cap T_j}\deg_{f^n}(x)\\ &\leq \frac{1}{d^n}\left(\sum_{j=0}^{n}\sum_{x\in M\cap T_j}D(d-1)^{n-j} + \sum_{x\in M\cap\widetilde{T}_n}D\right).\end{aligned}$$

Note that the more points in M lie in T_j with $j \in [0, n]$ as small as possible, the larger the right-hand side of the last inequality is. So the right-hand side of the last inequality is

$$\begin{aligned}
&\le\frac{1}{d^n}\left(\sum_{j=0}^{\lceil \log_d \lceil \frac{m}{t_0}\rceil\rceil} (\text{card } T_j)D(d-1)^{n-j} + mD\right)\\
&\le\frac{Dt_0}{d^n}\sum_{j=0}^{\lceil \log_d \lceil \frac{m}{t_0}\rceil\rceil} d^j(d-1)^{n-j} + \frac{mD}{d^n}\\
&\le Dt_0\left(\frac{d-1}{d}\right)^n \sum_{j=0}^{\lceil \log_d m\rceil}\left(\frac{d}{d-1}\right)^j + \frac{mD}{d^n}\\
&= Dt_0\left(\frac{d-1}{d}\right)^n \frac{\left(\frac{d}{d-1}\right)^{\lceil \log_d m\rceil+1}-1}{\frac{d}{d-1}-1} + \frac{mD}{d^n}\\
&\le Dt_0\left(\frac{d-1}{d}\right)^n \left(\frac{d}{d-1}\right)^{2+\log_d m}(d-1) + \frac{mD}{d^n}\\
&\le Dt_0\frac{d^2}{d-1}\left(\left(\frac{d-1}{d}\right)^{n-\log_d m} + \frac{m}{d^n}\right)\\
&= \frac{1}{2}E_f\left(d^{(n-\log_d m)\log_d \frac{d-1}{d}} + \frac{m}{d^n}\right)\\
&\le E_f \max\left\{\left(\frac{m}{d^n}\right)^{\log_d \frac{d}{d-1}}, \frac{m}{d^n}\right\},
\end{aligned}$$

where $E_f = 2Dt_0\frac{d^2}{d-1}$ is a constant that only depends on f. Thus in this case, $C = E_f$ and $\alpha = \log_d \frac{d}{d-1} \in (0, 1]$.

Case 3. Suppose that f has periodic critical points that may not be fixed points of f.

Set κ to be the product of the periods of all periodic critical points of f.

We claim that each periodic critical point of f^κ is a fixed point of f^κ. Indeed, if x is a periodic critical point of f^κ satisfying $f^{\kappa p}(x) = x$ for some $p \in \mathbb{N}$, then by (4.33), there exists an integer $i \in \{0, 1, \ldots, \kappa - 1\}$ such that $f^i(x) \in \text{crit } f$. Then $f^i(x)$ is a periodic critical point of f, so $f^\kappa(f^i(x)) = f^i(x)$. Thus

$$f^\kappa(x) = f^{\kappa-i}(f^i(x)) = f^{\kappa-i+\kappa}(f^i(x)) = \cdots = f^{\kappa-i+(p-1)\kappa}(f^i(x)) = f^{\kappa p}(x) = x.$$

The claim now follows.

Note that for each $n \in \mathbb{N}$,

$$\frac{1}{d^n}\sum_{x\in M}\deg_{f^n}(x) \le d^\kappa \frac{1}{d^{\kappa\lceil \frac{n}{\kappa}\rceil}}\sum_{x\in M}\deg_{f^{\kappa\lceil \frac{n}{\kappa}\rceil}}(x).$$

Hence by applying Case 2 for f^κ, we get a constant E_{f^κ} that depends only on f, such that the right-hand side of the above inequality is

$$\leq d^\kappa E_{f^\kappa} \max\left\{ \left(\frac{m}{d^{\kappa\lceil \frac{n}{\kappa}\rceil}}\right)^{\log_{d^\kappa}\frac{d^\kappa}{d^\kappa-1}}, \frac{m}{d^{\kappa\lceil\frac{n}{\kappa}\rceil}} \right\} \leq d^\kappa E_{f^\kappa} \max\left\{ \left(\frac{m}{d^n}\right)^{\log_{d^\kappa}\frac{d^\kappa}{d^\kappa-1}}, \frac{m}{d^n}\right\}.$$

Thus in this case $C = d^\kappa E_{f^\kappa}$ and $\alpha = \log_{d^\kappa} \frac{d^\kappa}{d^\kappa-1} \in (0, 1]$. □

Now we formulate a generalization of Lemma 4.21.

Lemma 4.28 *Let f be an expanding Thurston map, and $\mathscr{C} \subseteq S^2$ be an f^N-invariant Jordan curve containing* post f *for some $N \in \mathbb{N}$. Then there exists a constant $L \in [1, \deg f)$ with the following property:*

For each $m \in \mathbb{N}_0$, there exists a constant $D > 0$ such that for each $k \in \mathbb{N}_0$ and each m-edge e, there exists a collection M of $(m+k)$-tiles with card $M \leq DL^k$ *and $e \subseteq \operatorname{int}\Big(\bigcup_{X\in M} X\Big)$.*

Proof We denote $d = \deg f$, and consider the cell decompositions induced by $(f, \mathscr{C})$ in this proof.

Step 1: We first assume that for some $m \in \mathbb{N}$, there exist constants $L \in [1, d)$ and $D > 0$ such that for each $k \in \mathbb{N}_0$ and each m-edge e, there exists a collection M of $(m+k)$-tiles with card $M \leq DL^k$ and $e \subseteq \operatorname{int}\Big(\bigcup_{X\in M} X\Big)$. Then by Proposition 2.6(i), for each $(m-1)$-edge e, we can choose an m-edge e' such that $f(e') = e$. For each $k \in \mathbb{N}_0$, there exists a collection M' of $(m+k)$-tiles with card $M' \leq DL^k$ and $e' \subseteq \operatorname{int}\Big(\bigcup_{X\in M'} X\Big)$. We set M to be the collection $\{f(X) \mid X \in M'\}$ of $(m-1+k)$-tiles. Then card $M \leq$ card $M' \leq DL^k$ and $e \subseteq \operatorname{int}\Big(\bigcup_{X\in M} X\Big)$. Hence, it suffices to prove the lemma for "each $m \in \mathbb{N}_0$ with $m \equiv 0 \pmod N$" instead of "each $m \in \mathbb{N}_0$".

Step 2: We will prove the following statement by induction on κ:

For each $\kappa \in \{0, 1, \ldots, N-1\}$, there exists a constant $L_\kappa \in [1, d)$ with the following property:
For each $m \in \mathbb{N}_0$ with $m \equiv 0 \pmod N$, there exists a constant $D_\kappa > 0$ such that for each $k \in \mathbb{N}_0$ with $k \equiv \kappa \pmod N$ and each m-edge e, there exists a collection $M_{m,k,e}$ of $(m+k)$-tiles that satisfies card $M_{m,k,e} \leq D_\kappa L_\kappa^k$ and $e \subseteq \operatorname{int}\Big(\bigcup_{X\in M_{m,k,e}} X\Big)$.

Lemma 4.21 gives the case for $\kappa = 0$. For the induction step, we assume the above statement for some $\kappa \in [0, N-1]$.

Let $i \in \mathbb{N}_0$ and $p \in S^2$ be an i-vertex. We define the *i-flower* $W^i(p)$ as in [BM17] by

$$W^i(p) = \bigcup\{\operatorname{inte}(c) \mid c \in \mathbf{D}^i, p \in c\},$$

By definition, the only i-vertex contained in $W^i(p)$ is p, and the interior $\operatorname{inte}(e)$ of an i-edge e is a subset of $W^i(p)$ if and only if $p \in e$. Note that the number of i-tiles in $W^i(p)$ is $2\deg_{f^i}(p)$, i.e.,

$$\text{card}\,\{X \in \mathbf{X}^i \mid p \in X\} = 2\deg_{f^i}(p). \tag{4.34}$$

By [BM17, Lemma 5.37], there exists a constant $\beta \in \mathbb{N}$, which depends only on f and $\mathscr{C}$, such that for each $i \in \mathbb{N}$ and each i-tile $X \in \mathbf{X}^i$, X can be covered by a union of at most β $(i+1)$-flowers.

Fix an arbitrary $m \in \mathbb{N}_0$ with $m \equiv 0 \pmod N$, and fix an arbitrary m-edge e.

By the induction hypothesis, there exist constants $D_\kappa > 0$ and $L_\kappa \in [1, d)$ such that for each $k \in \mathbb{N}_0$ with $k \equiv \kappa + 1 \pmod N$, there exists a collection $M_{m,k-1,e}$ of $(m+k-1)$-tiles with $\text{card}\, M_{m,k-1,e} \leq D_\kappa L_\kappa^{k-1}$ and $e \subseteq \text{int}\Big(\bigcup_{X \in M_{m,k-1,e}} X\Big)$. Each $X \in M_{m,k-1,e}$ can be covered by β $(m+k)$-flowers $W^{m+k}(p)$. We can then construct a set $F \subseteq \mathbf{V}^{m+k}$ of $(m+k)$-vertices such that

$$\text{card}\, F \leq \beta D_\kappa L_\kappa^{k-1} \tag{4.35}$$

and

$$\bigcup_{X \in M_{m,k-1,e}} X \subseteq \bigcup_{p \in F} W^{m+k}(p). \tag{4.36}$$

We define

$$M_{m,k,e} = \{X \in \mathbf{X}^{m+k} \mid X \cap F \neq \emptyset\}. \tag{4.37}$$

Then $e \subseteq \text{int}\Big(\bigcup_{X \in M_{m,k,e}} X\Big)$, and by (4.34),

$$\text{card}\, M_{m,k,e} \leq \sum_{p \in F} 2\deg_{f^{m+k}}(p). \tag{4.38}$$

Since $L_\kappa \in [1, d)$, there exists $K \in \mathbb{N}$, depending only on f, $\mathscr{C}$, m, and κ, such that for each $i \geq K$, we have $\beta D_\kappa L_\kappa^{i-1} \leq d^{m+i}$.

Thus by (4.35), (4.38), and Lemma 4.27, for each $k \geq K$ with $k \equiv \kappa+1 \pmod N$, there exist constants $C > 0$ and $\alpha \in (0, 1]$, both of which depend only on f, such that

$$\begin{aligned} \text{card}\, M_{m,k,e} &\leq 2\sum_{p \in F} \deg_{f^{m+k}}(p) \leq 2Cd^{(m+k)(1-\alpha)} \left(\beta D_\kappa L_\kappa^{k-1}\right)^\alpha \\ &= 2Cd^{m(1-\alpha)} \beta^\alpha D_\kappa^\alpha L_\kappa^{-\alpha} \left(d^{1-\alpha} L_\kappa^\alpha\right)^k. \end{aligned} \tag{4.39}$$

Let $L_{\kappa+1} = d^{1-\alpha} L_\kappa^\alpha$. Since $L_\kappa \in [1, d)$, we get $L_{\kappa+1} \in [L_\kappa, d) \subseteq [1, d)$. Note that $L_{\kappa+1}$ only depends on f, $\mathscr{C}$, and κ. We define

$$\tau = \max\left\{2\sum_{p \in V} \deg_{f^{m+i}}(p) \,\middle|\, i \leq K,\, V \subseteq \mathbf{V}^{m+i},\, \text{card}\, V \leq \beta D_\kappa L_\kappa^{k-1}\right\}.$$

Since τ is the maximum over a finite set of numbers, $\tau < +\infty$. We set

$$D_{\kappa+1} = \max\{\tau, 2Cd^{m(1-\alpha)}\beta^{\alpha} D_{\kappa}^{\alpha} L_{\kappa}^{-\alpha}\}. \quad (4.40)$$

Then by (4.38), (4.39), and (4.40), we get that for each $k \in \mathbb{N}_0$ with $k \equiv \kappa + 1 \pmod N$,

$$\operatorname{card} M_{m,k,e} \le \sum_{p\in F} 2\deg_{f^{m+k}}(p) \le D_{\kappa+1} L_{\kappa+1}^{k}. \quad (4.41)$$

We note that τ only depends on f, $\mathscr{C}$, m, and κ, so $D_{\kappa+1}$ also only depends on f, $\mathscr{C}$, m, and κ.

This completes the induction.

Step 3: Now we define

$$L = \max\{L_{\kappa} \mid \kappa \in \{0, 1, \dots, N-1\}\}.$$

For each fixed $m \in \mathbb{N}_0$ with $m \equiv 0 \pmod N$, we set

$$D = \max\{D_{\kappa} \mid \kappa \in \{0, 1, \dots, N-1\}\},$$

and for each given $k \in \mathbb{N}_0$ and $e \in \mathbf{E}^m$, let $M = M_{m,k,e}$. Then we have $\operatorname{card} M \le DL^k$ and $e \subseteq \operatorname{int}\left(\bigcup_{X\in M} X\right)$. We note that here L only depends on f and $\mathscr{C}$, and on the other hand, D only depends on f, $\mathscr{C}$, and m. The proof is now complete. □

Remark 4.29 It is also possible to prove the previous lemma by observing that $\mathscr{C}$ equipped with the restriction of a visual metric d for f is a quasicircle (see [BM17, Theorem 15.3]), and S^2 equipped with d is linearly locally connected (see [BM17, Proposition 18.5]). A metric space X, that is homeomorphic to the plane and with $\overline{X}$ linearly locally connected and ∂X a Jordan curve, has the property that ∂X is porous in $\overline{X}$ (see [Wi07,Theorem IV.14]). Then we can mimic the original proof of Lemma 17.6 in [BM17].

We are finally ready to prove the equidistribution of the preperiodic points with respect to the measure of maximal entropy μ_f.

Proof (Proof of Theorem 4.3) Fix an arbitrary $N \ge N(f)$ where $N(f)$ is an constant as given in Corollary 2.18 depending only on f. We also fix an f^N-invariant Jordan curve $\mathscr{C}$ containing post f such that no N-tile in $\mathbf{D}^N(f, \mathscr{C})$ joins opposite sides of $\mathscr{C}$ as given in Corollary 2.18. In the proof below, we consider the cell decompositions $\mathbf{D}^i(f, \mathscr{C})$, $i \in \mathbb{N}_0$, induced by $(f, \mathscr{C})$, and denote $d = \deg f$.

Since ξ_n^m and $\widetilde{\xi}_n^m$ are Borel probability measures for all $m \in \mathbb{N}_0$ and $n \in \mathbb{N}$ with $m < n$, by Alaoglu's Theorem, it suffices to prove that in the weak* topology, every convergent subsequence of $\{\xi_n^{m_n}\}_{n\in\mathbb{N}}$ and $\{\widetilde{\xi}_n^{m_n}\}_{n\in\mathbb{N}}$ converges to μ_f.

Proof of (4.5):

Let $\{n_i\}_{i\in\mathbb{N}}$ be a strictly increasing sequence with

$$\xi_{n_i}^{m_{n_i}} \xrightarrow{w^*} \mu \text{ as } i \longrightarrow +\infty,$$

for some measure μ.

Case 1 for (4.5). We assume in this case that there is no constant $K \in \mathbb{N}$ such that for all $i \in \mathbb{N}$, $n_i - m_{n_i} \leq K$. Then by choosing a subsequence of $\{n_i\}_{i\in\mathbb{N}}$ if necessary, we can assume that $n_i - m_{n_i} \longrightarrow +\infty$ as $i \longrightarrow +\infty$.

Here is the idea of the proof in this case. By the spirit of Lemmas 4.6 and 4.7, there is an almost bijective correspondence between the fixed points of f^{n-m_n} and the $(n - m_n)$-tiles containing such points. The correspondence is particularly nice away from $\mathscr{C}$. Thus there is almost a bijective correspondence between the preperiodic points in $S_n^{m_n}$ and the n-tiles containing such points. So if we can control the behavior near $\mathscr{C}$, then Theorem 4.23 applies and we finish the proof in this case. Finally the control we need is provided by Lemma 4.28.

Now we start to implement this idea. We fix a 0-edge $e_0 \subseteq \mathscr{C}$. Observe that for each $i \in \mathbb{N}$, we can pair a white i-tile $X_w^i \in \mathbf{X}_w^i$ and a black i-tile $X_b^i \in \mathbf{X}_b^i$ whose intersection $X_w^i \cap X_b^i$ is an i-edge contained in $f^{-i}(e_0)$. There are a total of d^i such pairs and each i-tile is in exactly one such pair. We denote by $\mathbf{P}_i$ the collection of the unions $X_w^i \cup X_b^i$ of such pairs, i.e.,

$$\mathbf{P}_i = \{X_w^i \cup X_b^i \,|\, X_w^i \in \mathbf{X}_w^i, X_b^i \in \mathbf{X}_b^i, X_w^i \cap X_b^i \cap f^{-i}(e_0) \in \mathbf{E}^i\}.$$

We denote $\mathbf{P}_i' = \{A \in \mathbf{P}_i \,|\, A \cap \mathscr{C} = \emptyset\}$.

By Lemma 4.28, there exist $1 \leq L < d$ and $C > 0$ such that for each $i \in \mathbb{N}$ there exists a collection M of i-tiles with card $M \leq CL^i$ such that $\mathscr{C}$ is contained in the interior of the set $\bigcup_{X\in M} X$. Note that L and C are constants independent of i. Observe that for each $A \in \mathbf{P}_i$ that does not contain any i-tile in the collection M, we have $A \cap X' \subseteq \partial\Big(\bigcup_{X\in M} X\Big)$ for each $X' \in M$, so $A \cap \operatorname{int}\Big(\bigcup_{X\in M} X\Big) = \emptyset$. Since the number of distinct $A \in \mathbf{P}_i$ that contains an i-tile in M is bounded above by CL^i, we get

$$\operatorname{card}(\mathbf{P}_i') \geq d^i - CL^i. \tag{4.42}$$

Note that for each $i \in \mathbb{N}$ and each $A \in \mathbf{P}_i'$, either $A \subseteq X_w^0$ or $A \subseteq X_b^0$ where X_w^0 (resp. X_b^0) is the white (resp. black) 0-tile for $(f, \mathscr{C})$. So by Proposition 2.6(i) and Brouwer's Fixed Point Theorem, there is a map $\tau : \mathbf{P}_i' \to P_{1,f^i}$ from $\mathbf{P}_i'$ to the set of fixed points of f^i such that $\tau(A) \in A$. Note if a fixed point x of f^i has weight $\deg_{f^i}(x) > 1$, then x has to be contained in post $f \subseteq \mathscr{C}$. Thus $\deg_{f^i}(\tau(A)) = 1$ for all $A \in \mathbf{P}_i'$.

If for some $A \in \mathbf{P}_i'$, the point $\tau(A)$ were on the boundaries of the two i-tiles whose union is A, then $\tau(A)$ would have to be contained in $\mathscr{C}$ since the boundaries are mapped into $\mathscr{C}$ under f^i. Thus for each $A \in \mathbf{P}'_i$, the point $\tau(A)$ is contained in the interior of one of the two i-tiles whose union is A. Hence τ is injective. Moreover,

$$\deg_{f^{i+j}}(x) = 1 \text{ for each } j \in \mathbb{N}_0 \text{ and each } x \in \bigcup_{A \in \mathbf{P}'_i} f^{-j}(\tau(A)). \tag{4.43}$$

For each $i \in \mathbb{N}$, we choose a map $\beta_{n_i} : \mathbf{X}^{n_i}_w \to S^2$ by letting $\beta_{n_i}(X)$ be the unique point in $f^{-m_{n_i}}(\tau(A)) \cap B$ where $B \in \mathbf{P}_{n_i}$ with $X \subseteq B$, if there exists $A \in \mathbf{P}'_{n_i - m_{n_i}}$ with $f^{m_{n_i}}(X) \subseteq A$; and by letting $\beta_{n_i}(X)$ be an arbitrary point in X if there exists no $A \in \mathbf{P}'_{n_i - m_{n_i}}$ with $f^{m_{n_i}}(X) \subseteq A$.

We fix a visual metric d for f with expansion factor $\Lambda > 1$. Note that Λ depends only on f and d. Then $\operatorname{diam}_d(A) < c\Lambda^{-i}$ for each $i \in \mathbb{N}$, where $c \geq 1$ is a constant depending only on f, d, and $\mathscr{C}$ (see Lemma 2.13(ii)). Define $\alpha_n = c\Lambda^{-n}$ for each $n \in \mathbb{N}$. Thus α_{n_i} and β_{n_i} satisfy the hypothesis in Theorem 4.23. Define μ_{n_i} as in Theorem 4.23. Then

$$\mu_{n_i} \xrightarrow{w^*} \mu_f \text{ as } i \longrightarrow +\infty, \tag{4.44}$$

by Theorem 4.23.

We claim that the total variation $\left\| \mu_{n_i} - \xi^{m_{n_i}}_{n_i} \right\|$ of $\mu_{n_i} - \xi^{m_{n_i}}_{n_i}$ converges to 0 as $i \longrightarrow +\infty$.

Assuming the claim, then by (4.44), we can conclude that (4.5) holds in this case.

To prove the claim, by Corollary 4.14, we observe that for each $i \in \mathbb{N}$,

$$\begin{aligned} \left\| \mu_{n_i} - \xi^{m_{n_i}}_{n_i} \right\| \leq & \left\| \mu_{n_i} - \frac{1}{d^{n_i - m_{n_i}}} \sum_{A \in \mathbf{P}'_{n_i - m_{n_i}}} \frac{1}{d^{m_{n_i}}} \sum_{q \in f^{-m_{n_i}}(\tau(A))} \delta_q \right\| \\ & + \left\| \left(\frac{1}{d^{n_i}} - \frac{1}{d^{n_i} + d^{m_{n_i}}} \right) \sum_{A \in \mathbf{P}'_{n_i - m_{n_i}}} \sum_{q \in f^{-m_{n_i}}(\tau(A))} \delta_q \right\| \\ & + \left\| \frac{1}{d^{n_i - m_{n_i}} + 1} \sum_{A \in \mathbf{P}'_{n_i - m_{n_i}}} \frac{1}{d^{m_{n_i}}} \sum_{q \in f^{-m_{n_i}}(\tau(A))} \delta_q - \xi^{m_{n_i}}_{n_i} \right\|. \end{aligned} \tag{4.45}$$

In the first term on the right-hand side of (4.45), each δ_q in the summations cancels with the corresponding term in the definition of μ_{n_i}. So the first term on the right-hand side of (4.45) is equal to the difference of the total variations of the two measures, which by (4.42), is

$$\leq 1 - \frac{(d^{n_i - m_{n_i}} - CL^{n_i - m_{n_i}})d^{m_{n_i}}}{d^{n_i}} = C\left(\frac{L}{d}\right)^{n_i - m_{n_i}}.$$

In the second term on the right-hand side of (4.45), the total number of terms in the summations is bounded above by d^{n_i}. So the second term on the right-hand side of (4.45) is

$$\leq \left| \frac{1}{d^{n_i}} - \frac{1}{d^{n_i} + d^{m_i}} \right| d^{n_i}.$$

In the third term on the right-hand side of (4.45), by (4.43), $\deg_{f^{n_i}}(q) = 1$ for each $A \in \mathbf{P}'_{n_i - m_{n_i}}$ and each $q \in f^{-m_{n_i}}(\tau(A))$. So by (4.4) and Corollary 4.14, each δ_q in the summations cancels with the corresponding δ_q in $\xi_{n_i}^{m_{n_i}}$. So the third term on the right-hand side of (4.45) is equal to the difference of the total variations of the two measures, which by (4.42) and Corollary 4.14, is

$$\leq 1 - \frac{(d^{n_i - m_{n_i}} - CL^{n_i - m_{n_i}})d^{m_{n_i}}}{(d^{n_i - m_{n_i}} + 1)d^{m_{n_i}}} = \frac{1 + CL^{n_i - m_{n_i}}}{d^{n_i - m_{n_i}} + 1}.$$

Since $n_i - m_{n_i} \longrightarrow +\infty$ as $i \longrightarrow +\infty$, each term on the right-hand side of (4.45) converges to 0 as $i \longrightarrow +\infty$. So

$$\left\| \mu_{n_i} - \xi_{n_i}^{m_{n_i}} \right\| \longrightarrow 0 \text{ as } i \longrightarrow +\infty$$

as claimed.

Case 2 for (4.5). We assume in this case that there is a constant $K \in \mathbb{N}$ such that for all $i \in \mathbb{N}$, $n_i - m_{n_i} \leq K$. Then by choosing a subsequence of $\{n_i\}_{i \in \mathbb{N}}$ if necessary, we can assume that there exists some constant $l \in [0, K]$ such that for all $i \in \mathbb{N}$, $n_i - m_{n_i} = l$. Note that in this case, $m_{n_i} \longrightarrow +\infty$ as $i \longrightarrow +\infty$.

Then by Corollary 4.14 and Theorem 4.1,

$$\begin{aligned} \xi_{n_i}^{m_{n_i}} &= \frac{1}{d^{m_{n_i}}(d^l + 1)} \sum_{x \in S_{n_i}^{m_{n_i}}} \deg_{f^{n_i}}(x)\delta_x \\ &= \frac{1}{d^l + 1} \sum_{y = f^l(y)} \deg_{f^l}(y) \left(\frac{1}{d^{m_{n_i}}} \sum_{x \in f^{-m_{n_i}}(y)} \deg_{f^{m_{n_i}}}(x)\delta_x \right). \end{aligned}$$

By Theorem 4.2, for each $y \in S^2$,

$$\frac{1}{d^{m_{n_i}}} \sum_{x \in f^{-m_{n_i}}(y)} \deg_{f^{m_{n_i}}}(x)\delta_x \xrightarrow{w^*} \mu_f \text{ as } i \longrightarrow +\infty.$$

So each term in the sequence $\{\xi_{n_i}^{m_{n_i}}\}_{i \in \mathbb{N}}$ is a convex combination of the corresponding terms in sequences of measures, each of which converges in the weak* topology to μ_f. Hence by Lemma 4.18, $\{\xi_{n_i}^{m_{n_i}}\}_{i \in \mathbb{N}}$ also converges to μ_f in the weak* topology. It then follows that $\mu = \mu_f$. Thus (4.5) follows in this case.

Proof of (4.6):

Let $\{n_i\}_{i \in \mathbb{N}}$ be a strictly increasing sequence with

$$\widetilde{\xi}_{n_i}^{m_{n_i}} \xrightarrow{w^*} \widetilde{\mu} \text{ as } i \longrightarrow +\infty,$$

for some measure $\widetilde{\mu}$.

Case 1 for (4.6). We assume in this case that there is no constant $K \in \mathbb{N}$ such that for all $i \in \mathbb{N}$, $n_i - m_{n_i} \leq K$. Then by choosing a subsequence of $\{n_i\}_{i \in \mathbb{N}}$ if necessary, we can assume that $n_i - m_{n_i} \longrightarrow +\infty$ as $i \longrightarrow +\infty$.

The idea of the proof in this case is similar to that of the proof of Case 1 for (4.5). We use the same notation as in the proof of Case 1 for (4.5). Then (4.6) follows in this case if we can prove that $\left\| \mu_{n_i} - \widetilde{\xi}_{n_i}^{m_{n_i}} \right\|$ converges to 0 as $i \longrightarrow +\infty$.

As before, we observe that

$$\begin{aligned}\left\| \mu_{n_i} - \widetilde{\xi}_{n_i}^{m_{n_i}} \right\| \leq & \left\| \mu_{n_i} - \frac{1}{d^{n_i - m_{n_i}}} \sum_{A \in \mathbf{P}'_{n_i - m_{n_i}}} \frac{1}{d^{m_{n_i}}} \sum_{q \in f^{-m_{n_i}}(\tau(A))} \delta_q \right\| \\ & + \left\| \left(\frac{1}{d^{n_i}} - \frac{1}{\widetilde{s}_{n_i}^{m_{n_i}}} \right) \sum_{A \in \mathbf{P}'_{n_i - m_{n_i}}} \sum_{q \in f^{-m_{n_i}}(\tau(A))} \delta_q \right\| \\ & + \left\| \frac{1}{\widetilde{s}_{n_i}^{m_{n_i}}} \sum_{A \in \mathbf{P}'_{n_i - m_{n_i}}} \sum_{q \in f^{-m_{n_i}}(\tau(A))} \delta_q - \widetilde{\xi}_{n_i}^{m_{n_i}} \right\|.\end{aligned} \tag{4.46}$$

As the first term on the right-hand side of (4.45) discussed before, the first term on the right-hand side of (4.46) is

$$\leq 1 - \frac{(d^{n_i - m_{n_i}} - CL^{n_i - m_{n_i}}) d^{m_{n_i}}}{d^{n_i}} = C \left(\frac{L}{d} \right)^{n_i - m_{n_i}}.$$

In the second term on the right-hand side of (4.46), the total number of terms in the summations is bounded above by d^{n_i}. By (4.42), (4.43), and Corollary 4.14, we have

$$d^{m_{n_i}} (d^{n_i - m_{n_i}} + 1) = s_{n_i}^{m_{n_i}} \geq \widetilde{s}_{n_i}^{m_{n_i}} \geq d^{m_{n_i}} \operatorname{card}(\mathbf{P}'_{n_i - m_{n_i}}) \geq d^{m_{n_i}} (d^{n_i - m_{n_i}} - CL^{n_i - m_{n_i}}). \tag{4.47}$$

So the second term on the right-hand side of (4.46) is

$$\leq \left| \frac{1}{d^{n_i}} - \frac{1}{\widetilde{s}_{n_i}^{m_{n_i}}} \right| d^{n_i} = \left| 1 - \frac{d^{n_i}}{\widetilde{s}_{n_i}^{m_{n_i}}} \right| \leq \max \left\{ \frac{1}{d^{n_i - m_{n_i}}}, \frac{CL^{n_i - m_{n_i}}}{d^{n_i - m_{n_i}}} \right\}.$$

In the third term on the right-hand side of (4.46), by (4.43), $\deg_{f^{n_i}}(q) = 1$ for each $A \in \mathbf{P}'_{n_i - m_{n_i}}$ and each $q \in f^{-m_{n_i}}(\tau(A))$. So by (4.4), each δ_q in the summations cancels with the corresponding δ_q in $\widetilde{\xi}_{n_i}^{m_{n_i}}$. So the third term on the right-hand side of (4.46) is equal to the difference of the total variations of the two measures, which by (4.47) and (4.42), for $n_i - m_{n_i}$ large enough, is

$$\leq \frac{d^{n_i} + d^{m_{n_i}} - (d^{n_i - m_{n_i}} - CL^{n_i - m_{n_i}})d^{m_{n_i}}}{\widetilde{S}_{n_i}^{m_{n_i}}} \leq \frac{1 + CL^{n_i - m_{n_i}}}{d^{n_i - m_{n_i}} - CL^{n_i - m_{n_i}}}.$$

Since $n_i - m_{n_i} \longrightarrow +\infty$ as $i \longrightarrow +\infty$, each term on the right-hand side of (4.46) converges to 0 as $i \longrightarrow +\infty$. So we can conclude that

$$\left\| \mu_{n_i} - \widetilde{\xi}_{n_i}^{m_{n_i}} \right\| \longrightarrow 0 \text{ as } i \longrightarrow +\infty.$$

So $\widetilde{\mu} = \mu_f$. Thus (4.6) follows in this case.

Case 2 for (4.6). We assume in this case that there is a constant $K \in \mathbb{N}$ such that for all $i \in \mathbb{N}$, $n_i - m_{n_i} \leq K$. Then by choosing a subsequence of $\{n_i\}_{i \in \mathbb{N}}$ if necessary, we can assume that there exists some constant $l \in [0, K]$ such that for all $i \in \mathbb{N}$, $n_i - m_{n_i} = l$. Note that in this case, $m_{n_i} \longrightarrow +\infty$ as $i \longrightarrow +\infty$.

Then for each $i \in \mathbb{N}$, we have

$$\widetilde{\xi}_{n_i}^{m_{n_i}} = \frac{1}{\widetilde{S}_{n_i}^{m_{n_i}}} \sum_{x \in \widetilde{S}_{n_i}^{m_{n_i}}} \delta_x = \frac{1}{\widetilde{S}_{n_i}^{m_{n_i}}} \sum_{y = f^l(y)} Z_{m_{n_i}, y} \left(\frac{1}{Z_{m_{n_i}, y}} \sum_{x \in f^{-m_{n_i}}(y)} \delta_x \right),$$

where $Z_{m,y} = \operatorname{card}\left(f^{-m}(y)\right)$ for each $y \in S^2$ and each $m \in \mathbb{N}_0$. Note that for each $i \in \mathbb{N}$, we have

$$\widetilde{S}_{n_i}^{m_{n_i}} = \sum_{y = f^l(y)} Z_{m_{n_i}, y}.$$

Denote, for each $i \in \mathbb{N}$ and each $y \in S^2$, the Borel probability measure $\mu_{i,y} = \frac{1}{Z_{m_{n_i}, y}} \sum_{x \in f^{-m_{n_i}}(y)} \delta_x$. Then by Theorem 4.2, we have

$$\mu_{i,y} \xrightarrow{w^*} \mu_f \text{ as } i \longrightarrow +\infty.$$

So each term in $\{\widetilde{\xi}_{n_i}^{m_{n_i}}\}_{i \in \mathbb{N}}$ is a convex combination of the corresponding terms in sequences of measures, each of which converges in the weak* topology to μ_f. Hence by Lemma 4.18, $\{\widetilde{\xi}_{n_i}^{m_{n_i}}\}_{i \in \mathbb{N}}$ also converges to μ_f in the weak* topology. It then follows that $\widetilde{\mu} = \mu_f$. Thus (4.6) follows in this case. □

The proof of Theorem 4.3 also gives us the following corollary.

Corollary 4.30 *Let f be an expanding Thurston map. If $\{m_n\}_{n \in \mathbb{N}}$ is a sequence in $\mathbb{N}_0$ such that $m_n < n$ for each $n \in \mathbb{N}$ and $\lim_{n \to +\infty} n - m_n = +\infty$, then*

$$\lim_{n \to +\infty} \frac{\widetilde{S}_n^{m_n}}{S_n^{m_n}} = 1. \tag{4.48}$$

Proof By the proof of Theorem 4.3, especially (4.47), we get that for each $n \in \mathbb{N}$,

$$\frac{d^{n-m_n} - CL^{n-m_n}}{d^{n-m_n} + 1} \leq \frac{\widetilde{s}_n^{m_n}}{s_n^{m_n}} \leq 1, \tag{4.49}$$

where $d = \deg f$. Then (4.48) follows from the fact that $1 \leq L < d$ and the condition that $\lim\limits_{n \to +\infty} n - m_n = +\infty$. □

By (4.23), Theorem 4.1, and Corollary 4.30 with $m_n = 0$ for each $n \in \mathbb{N}$, we get the following corollary, which is an analog of the corresponding result for expansive homeomorphisms on compact metric spaces with the *specification property* (see for example, [KH95, Theorem 18.5.5]).

Corollary 4.31 *Let f be an expanding Thurston map. Then for each constant $c \in (0, 1)$, there exists a constant $N \in \mathbb{N}$ such that for each $n \geq N$,*

$$\begin{aligned} ce^{nh_{\rm top}(f)} = c(\deg f)^n &< \operatorname{card}\{x \in S^2 \,|\, f^n(x) = x\} \\ &\leq \sum_{x = f^n(x)} \deg_{f^n}(x) = (\deg f)^n + 1 < \frac{1}{c} e^{nh_{\rm top}(f)}. \end{aligned}$$

In particular,

$$\lim_{n \to +\infty} \frac{\operatorname{card}\{x \in S^2 \,|\, f^n(x) = x\}}{\exp\left(nh_{\rm top}(f)\right)} = \lim_{n \to +\infty} \frac{\operatorname{card}\{x \in S^2 \,|\, f^n(x) = x\}}{(\deg f)^n} = 1.$$

Finally, we get the equidistribution of the periodic points with respect to the measure of maximal entropy μ_f as an immediate corollary.

Proof (Proof of Corollary 4.4) We get (4.7) and (4.8) from Theorem 4.3 with $m_n = 0$ for all $n \in \mathbb{N}$. Then (4.9) follows from (4.8) and Corollary 4.31. □

4.3 Expanding Thurston Maps as Factors of the Left-Shift

M. Bonk and D. Meyer [BM17] proved that for an expanding Thurston map f, the topological dynamical system (S^2, f) is a factor of a certain classical topological dynamical system, namely, the left-shift on the one-sided infinite sequences of $\deg f$ symbols. The goal of this section is to generalize this result to the category of measure-preserving dynamical systems. The invariant measure for each measure-preserving dynamical system considered in this section is going to be the unique measure of maximal entropy of the corresponding system.

Let X and $\widetilde{X}$ be topological spaces, and $f: X \to X$ and $\widetilde{f}: \widetilde{X} \to \widetilde{X}$ be continuous maps. We say that the topological dynamical system (X, f) is a *factor of the topological dynamical system* $(\widetilde{X}, \widetilde{f})$ if there is a surjective continuous map $\varphi: \widetilde{X} \to X$ such that $\varphi \circ \widetilde{f} = f \circ \varphi$. For measure-preserving dynamical systems (X, g, μ) and $(\widetilde{X}, \widetilde{g}, \widetilde{\mu})$ where X and $\widetilde{X}$ are measure spaces, $g: X \to X$ and $\widetilde{g}: \widetilde{X} \to \widetilde{X}$ measurable maps, and $\mu \in \mathcal{M}(X, g)$ and $\widetilde{\mu} \in \mathcal{M}(\widetilde{X}, \widetilde{g})$, we say that the measure-preserving dynamical system $(\widetilde{X}, \widetilde{g}, \widetilde{\mu})$ is a *factor of the measure-preserving dynamical system* (X, g, μ) if there is a measurable map $\varphi : \widetilde{X} \to X$ such that $\varphi \circ \widetilde{g} = g \circ \varphi$ and $\varphi_* \widetilde{\mu} = \mu$. Thus we get the following commutative diagram:

$$\begin{array}{ccc} \widetilde{X} & \xrightarrow{\widetilde{f}} & \widetilde{X} \\ {\scriptstyle\varphi}\downarrow & & \downarrow{\scriptstyle\varphi} \\ X & \xrightarrow{f} & X \end{array}$$

We recall a classical example of symbolic dynamical systems, namely (J_k^ω, Σ), where the *alphabet* $J_k = \{0, 1, \dots, k-1\}$ for some $k \in \mathbb{N}$, the *set of infinite words* $J_k^\omega = \prod\limits_{i=1}^{+\infty} J_k$, and Σ is the left-shift operator with

$$\Sigma(i_1, i_2, \dots) = (i_2, i_3, \dots)$$

for each $(i_i, i_2, \dots) \in J_k^\omega$. We equip J_k^ω with a metric d such that the distance between two distinct infinite words $(i_1, i_2, \dots)$ and $(j_1, j_2, \dots)$ is $\frac{1}{m}$, where $m = \min\{n \in \mathbb{N} \mid i_n \neq j_n\}$.

Define the *set of words of length n* as $J_k^n = \prod_{i=1}^n J_k$, for $n \in \mathbb{N}$ and $J_k^0 = \{\emptyset\}$ where $\emptyset$ is considered as the word of length 0, which is also denoted by (). Denote the *set of finite words* by $J_k^* = \bigcup\limits_{n=0}^{+\infty} J_k^n$. Then the left-shift operator Σ is defined on $J_k^* \backslash J_k^0$ naturally by

$$\Sigma(i_1, i_2, \dots, i_n) = (i_2, i_3, \dots, i_n).$$

It is well-known that the dynamical system (J_k^ω, Σ) has a unique measure of maximal entropy η_Σ, which is characterized by the property that

$$\eta_\Sigma\left(C(j_1, j_2, \dots, j_n)\right) = k^{-n},$$

for $n \in \mathbb{N}$ and $j_1, j_2, \dots, j_n \in J_k$, where

$$C(j_1, j_2, \dots, j_n) = \{(i_1, i_2, \dots) \in J_k^\omega \mid i_1 = j_1, i_2 = j_2, \dots, i_n = j_n\} \tag{4.50}$$

is the *cylinder set* determined by $j_1, j_2, \dots, j_n$ (see for example, [KH95, Sect. 4.4]).

We will prove that for each expanding Thurston map f with $\deg f = k$ and its measure of maximal entropy μ_f, the measure-preserving dynamical system (S^2, f, μ_f) is a factor of the system $(J_k^\omega, \Sigma, \eta_\Sigma)$.

We now review a construction from [BM17] for the convenience of the reader.

Let $f : S^2 \to S^2$ be an expanding Thurston map, and $\mathcal{C} \subseteq S^2$ a Jordan curve with $\operatorname{post} f \subseteq \mathcal{C}$. Consider the cell decompositions induced by the pair $(f, \mathcal{C})$. Let $k = \deg f$. Fix an arbitrary point $p \in \operatorname{inte}(X_w^0)$. Let $q_1, q_2, \dots, q_k$ be the distinct points in $f^{-1}(p)$. For $i = 1, 2, \dots, k$, we pick a continuous path $\alpha_i : [0, 1] \to S^2 \setminus \operatorname{post} f$ with $\alpha_i(0) = p$ and $\alpha_i(1) = q_i$.

We construct $\psi : J_k^* \to S^2$ inductively such that $\psi(I) \in f^{-n}(p)$, for each $n \in \mathbb{N}_0$ and $I \in J_k^n$, in the following way:

Define $\psi(\emptyset) = p$, and $\psi((i)) = q_i$ for each $(i) \in J_k^1$. Suppose that ψ has been defined for all $I \in \bigcup_{j=0}^{n} J_k^j$, where $n \in \mathbb{N}$. Now for each $(i_1, i_2, \dots, i_{n+1}) \in J_k^{n+1}$, the point $\psi((i_1, i_2, \dots, i_n)) \in f^{-n}(p)$ has already been defined. Since $f^n(\psi((i_1, i_2, \dots, i_n))) = p$ and $f^n : S^2 \setminus f^{-n}(\operatorname{post} f) \to S^2 \setminus \operatorname{post} f$ is a covering map, the path $\alpha_{i_{n+1}}$ has a unique lift $\widetilde{\alpha}_{i_{n+1}} : [0, 1] \to S^2$ with $\widetilde{\alpha}_{i_{n+1}}(0) = \psi((i_1, i_2, \dots, i_n))$ and $f^n \circ \widetilde{\alpha}_{i_{n+1}} = \alpha_{i_{n+1}}$. We now define $\psi((i_1, i_2, \dots, i_{n+1})) = \widetilde{\alpha}_{i_{n+1}}(1)$. Note that then

$$f^{n+1}(\psi((i_1, i_2, \dots, i_{n+1}))) = f^{n+1}(\widetilde{\alpha}_{i_{n+1}}(1)) = f(\alpha_{i_{n+1}}(1)) = f(q_{i_{n+1}}) = p.$$

Hence $\psi((i_1, i_2, \dots, i_{n+1})) \in f^{-(n+1)}(p)$. This completes the inductive construction of ψ.

Note that $\psi : J_k^* \to S^2$ induces a map $\widetilde{\psi} : J_k^* \to \bigcup_{n=0}^{+\infty} \mathbf{X}_w^n$ by mapping each $(i_1, i_2, \dots, i_n) \in J_k^n$ to the unique white n-tile $X_w^n \in \mathbf{X}_w^n$ containing $\psi((i_1, i_2, \dots, i_n)) \in f^{-n}(p)$.

By the proof of Theorem 9.1 of [BM17], for each $n \in \mathbb{N}$, $\psi|_{J_k^n} : J_k^n \to f^{-n}(p)$ is a bijection. Hence $\widetilde{\psi}|_{J_k^n} : J_k^n \to \mathbf{X}_w^n$ for $n \in \mathbb{N}_0$, and $\widetilde{\psi} : J_k^* \to \bigcup_{n=0}^{+\infty} \mathbf{X}_w^n$ are also bijections. Moreover, by the proof of Theorem 9.1 in [BM17], we have that for each $(i_1, i_2, \dots) \in J_k^\omega$, $\{\psi((i_1, i_2, \dots, i_n))\}_{n \in \mathbb{N}}$ is a Cauchy sequence in (S^2, d), for each visual metric d for f. So as shown in the proof of Theorem 9.1 in [BM17], the map $\varphi : J_k^\omega \to S^2$ defined by

$$\varphi((i_1, i_2, \dots)) = \lim_{n \to +\infty} \psi((i_1, i_2, \dots, i_n)) \tag{4.51}$$

satisfies

1. φ is continuous,
2. $f \circ \varphi = \varphi \circ \Sigma$,
3. $\varphi : J_k^\omega \to S^2$ is surjective.

We now reformulate Theorem 9.1 from [BM17] in the following way.

Theorem 4.32 (M. Bonk and D. Meyer) *Let $f : S^2 \to S^2$ be an expanding Thurston map with $\deg f = k$. Then (S^2, f) is a factor of the topological dynamical system (J_k^ω, Σ). More precisely, the surjective continuous map $\varphi : J_k^\omega \to S^2$ defined above satisfies $f \circ \varphi = \varphi \circ \Sigma$.*

We will strengthen Theorem 4.32 in the following theorem.

Theorem 4.33 *Let $f: S^2 \to S^2$ be an expanding Thurston map with $\deg f = k$. Then (S^2, f, μ_f) is a factor of the measure-preserving dynamical system $(J_k^\omega, \Sigma, \eta_\Sigma)$, where μ_f and η_Σ are the unique measures of maximal entropy of (S^2, f) and (J_k^ω, Σ), respectively. More precisely, the surjective continuous map $\varphi: J_k^\omega \to S^2$ defined above satisfies $f \circ \varphi = \varphi \circ \Sigma$ and $\varphi_* \eta_\Sigma = \mu_f$.*

Proof Let $\mathscr{C} \subseteq S^2$ be a Jordan curve containing post f. Let d be a visual metric on S^2 for f with expansion factor $\Lambda > 1$. Note that Λ depends only on f and d. Consider the cell decompositions induced by $(f, \mathscr{C})$.

By Theorem 4.32, it suffices to prove that $\varphi_* \eta_\Sigma = \mu_f$.

For each $n \in \mathbb{N}$, we fix a function $\widetilde{\beta}_n : J_k^n \to J_k^\omega$ which maps each $(i_1, i_2, \dots, i_n) \in J_k^n$ to $(i_1, i_2, \dots, i_n, i_{n+1}, \dots) \in J_k^\omega$, for some arbitrarily chosen $i_{n+1}, i_{n+2}, \dots \in J_k$ depending on $i_1, i_2, \dots, i_n$. In other words, $\widetilde{\beta}_n$ extends a finite word of length n to an arbitrary infinite word.

Define $\beta_n = \varphi \circ \widetilde{\beta}_n \circ \widetilde{\psi}^{-1}$, for each $n \in \mathbb{N}$, where $\widetilde{\psi}$ is defined earlier in this section.

We claim that the maps $\beta_n : \mathbf{X}_w^n \to S^2$ with $n \in \mathbb{N}$ satisfy the hypothesis for β_n in Theorem 4.23, namely,

$$\max\{d(\beta_n(X_w^n), X_w^n) \mid X_w^n \in \mathbf{X}_w^n\} \longrightarrow 0 \text{ as } n \longrightarrow +\infty.$$

Indeed, by the construction of φ, $\widetilde{\beta}_n$, ψ and $\widetilde{\psi}$ above, we have that β_n maps a white n-tile X_w^n to the limit of a Cauchy sequence

$$(\psi((j_1, j_2, \dots, j_m)))_{m \in \mathbb{N}}$$

such that $\psi((j_1, j_2, \dots, j_n)) \in X_w^n$. Since for each $m \in \mathbb{N}$, the points $\psi((j_1, j_2, \dots, j_m))$ and $\psi((j_1, j_2, \dots, j_{m+1}))$ are joined by a lift of one of the paths $\alpha_1, \alpha_2, \dots, \alpha_k$ (defined above) by f^m, by Lemma 8.9 in [BM17], we have that

$$d\left(\psi((j_1, j_2, \dots, j_m)), \psi((j_1, j_2, \dots, j_{m+1}))\right) \leq C\Lambda^{-m},$$

for all $m \in \mathbb{N}$, where $C > 0$ is a constant depending only on f, d, and the curves α_i, $i \in \{1, 2, \dots, k\}$, in the construction of ψ. In particular, both C and Λ are independent of m and $(j_1, j_2, \dots) \in J_k^\omega$. So $d(\beta_n(X_w^n), X_w^n) \leq C\frac{\Lambda^n}{1-\Lambda}$ for each $n \in \mathbb{N}$ and each $X_w^n \in \mathbf{X}_w^n$. The above claim follows.

For $i \in \mathbb{N}$, define

$$\eta_i = \frac{1}{k^i} \sum_{I \in J_k^i} \delta_{\widetilde{\beta}_i(I)}.$$

Observe that for all $n \in \mathbb{N}$ and $m \in \mathbb{N}$ with $m \geq n$, and each $(i_1, i_2, \dots, i_n) \in J_k^n$, we have

$$\eta_m(C(i_1, i_2, \dots, i_n)) = \eta_\Sigma(C(i_1, i_2, \dots, i_n)),$$

where $C(i_1, i_2, \ldots, i_n)$ is defined in (4.50). So by the uniform continuity of each continuous function on J_k^ω, it is easy to see that

$$\eta_i \xrightarrow{w^*} \eta_\Sigma \text{ as } i \longrightarrow +\infty. \tag{4.52}$$

Note that since $\widetilde{\psi}|_{J_k^n} : J_k^n \to \mathbf{X}_w^n$ is a bijection for each $n \in \mathbb{N}_0$, we have for each $i \in \mathbb{N}$,

$$\varphi_* \eta_i = \frac{1}{k^i} \sum_{I \in J_k^i} \delta_{\varphi \circ \widetilde{\beta}_i(I)} = \frac{1}{k^i} \sum_{X^i \in \mathbf{X}_w^i} \delta_{\varphi \circ \widetilde{\beta}_i \circ \widetilde{\psi}^{-1}(X^i)} = \frac{1}{k^i} \sum_{X^i \in \mathbf{X}_w^i} \delta_{\beta_i(X^i)}.$$

Hence, by Theorem 4.23,

$$\varphi_* \eta_i \xrightarrow{w^*} \mu_f \text{ as } i \longrightarrow +\infty. \tag{4.53}$$

Therefore, by (4.52), (4.53), and Lemma 4.17, we can conclude that $\varphi_* \eta_\Sigma = \mu_f$. □

Chapter 5
Equilibrium States

In this chapter, we investigate the existence, uniqueness, and other properties of equilibrium states for an expanding Thurston map. The main tool for this chapter is the thermodynamical formalism. We record the main theorem of this chapter below.

Theorem 5.1 *Let $f\colon S^2 \rightarrow S^2$ be an expanding Thurston map and d be a visual metric on S^2 for f. Let ϕ be a real-valued Hölder continuous function on S^2 with respect to the metric d.*

Then there exists a unique equilibrium state μ_ϕ for the map f and the potential ϕ. If ψ is another real-valued Hölder continuous function on S^2 with respect to the metric d, then $\mu_\phi = \mu_\psi$ if and only if there exists a constant $K \in \mathbb{R}$ such that $\phi - \psi$ and $K \mathbb{1}_{S^2}$ are co-homologous in the space of real-valued continuous functions on S^2, i.e., $\phi - \psi - K \mathbb{1}_{S^2} = u \circ f - u$ for some real-valued continuous function u on S^2.

Moreover, μ_ϕ is a non-atomic f-invariant Borel probability measure on S^2 and the measure-preserving transformation f of the probability space (S^2, μ_ϕ) is forward quasi-invariant, nonsingular, exact, and in particular, mixing and ergodic.

In addition, the preimages points of f are equidistributed with respect to μ_ϕ, i.e., for each sequence $\{x_n\}_{n\in\mathbb{N}}$ of points in S^2, as $n \longrightarrow +\infty$,

$$\frac{1}{Z_n(\phi)} \sum_{y \in f^{-n}(x_n)} \deg_{f^n}(y) \exp\left(S_n\phi(y)\right) \frac{1}{n} \sum_{i=0}^{n-1} \delta_{f^i(y)} \xrightarrow{w^*} \mu_\phi, \tag{5.1}$$

$$\frac{1}{Z_n\big(\widetilde{\phi}\big)} \sum_{y \in f^{-n}(x_n)} \deg_{f^n}(y) \exp\big(S_n\widetilde{\phi}(y)\big) \delta_y \xrightarrow{w^*} \mu_\phi, \tag{5.2}$$

where $Z_n(\psi) = \sum\limits_{y \in f^{-n}(x_n)} \deg_{f^n}(y) \exp\left(S_n\psi(y)\right)$, for each $n \in \mathbb{N}$ and each $\psi \in C(S^2)$.

Z. Li, *Ergodic Theory of Expanding Thurston Maps*, Atlantis Studies in Dynamical Systems 4, DOI 10.2991/978-94-6239-174-1_5

Here the symbol w^* indicates convergence in the weak* topology, $\deg_{f^n}(x)$ denotes the local degree of the map f^n at x, $S_n\psi(y) = \sum_{i=0}^{n-1} \psi(f^i(y))$, and $\widetilde{\phi}$ is a potential related to ϕ defined in (5.48).

This theorem combines Theorem 5.36, Theorem 5.41, Corollary 5.42, Corollary 5.44, Theorem 5.45, and Proposition 5.54.

As a quick consequence of the proof of the uniqueness of the equilibrium state, we show in Proposition 5.38 that under the assumptions in Theorem 5.1, the images of each Borel probability measure μ under iterates of the adjoint of the Ruelle operator $\mathcal{L}_{\widetilde{\phi}}$ converge in the weak* topology to the unique equilibrium state μ_ϕ, i.e.,

$$\left(\mathcal{L}^*_{\widetilde{\phi}}\right)^n(\mu) \xrightarrow{w^*} \mu_\phi \text{ as } n \longrightarrow +\infty. \tag{5.3}$$

A rational Thurston map is expanding if and only if it has no periodic critical points (see [BM17, Proposition 2.3]). So when we restrict to rational Thurston maps, we get the following corollary as an immediate consequence of Theorem 5.1 and Remark 2.14.

Corollary 5.2 *Let f be a postcritically-finite rational map on the Riemann sphere $\widehat{\mathbb{C}}$ with no periodic critical points and with degree at least 2. Let ϕ be a real-valued Hölder continuous function on $\widehat{\mathbb{C}}$ equipped with the chordal metric.*

Then there exists a unique equilibrium state μ_ϕ for the map f and the potential ϕ. If ψ is another real-valued Hölder continuous function on $\widehat{\mathbb{C}}$, then $\mu_\phi = \mu_\psi$ if and only if there exists a constant $K \in \mathbb{R}$ such that $\phi - \psi$ and $K\mathbb{1}_{\widehat{\mathbb{C}}}$ are co-homologous in the space of real-valued continuous functions on $\widehat{\mathbb{C}}$, i.e., $\phi - \psi - K\mathbb{1}_{\widehat{\mathbb{C}}} = u \circ f - u$ for some real-valued continuous function u on $\widehat{\mathbb{C}}$.

Moreover, μ_ϕ is a non-atomic f-invariant Borel probability measure on S^2 and the measure-preserving transformation f of the probability space (S^2, μ_ϕ) is forward quasi-invariant, nonsingular, exact, and in particular, mixing and ergodic.

In addition, both (5.1) and (5.2) hold as $n \longrightarrow +\infty$.

The expression "postcritically-finite rational map (with degree at least 2)" is another name for a rational Thurston map, used by many authors in holomorphic dynamics.

In Sect. 5.1, we state the assumptions on some of the objects in the remaining part of this monograph, which we are going to repeatedly refer to later as *the Assumptions*.

In Sect. 5.2, following the ideas from [PU10] and [Zi96], we use the thermodynamical formalism to prove the existence of the equilibrium states for expanding Thurston maps and real-valued Hölder continuous potentials. We first establish two distortion lemmas (Lemmas 5.3 and 5.4), which will be used frequently throughout this paper. Next, we define *Gibbs states* and *radial Gibbs states*. Later in Proposition 5.21, we prove that for an expanding Thurston map the notion of a Gibbs state is equivalent to that of a radial Gibbs state if and only if the map does not have periodic critical points.

By applying the Schauder-Tikhonov Fixed Point Theorem, we establish in Theorem 5.12 the existence of an eigenmeasure m_ϕ of the adjoint $\mathscr{L}_\phi^*$ of the Ruelle operator $\mathscr{L}_\phi$, for a real-valued Hölder continuous potential ϕ. We also show in Theorem 5.12 that the Jacobian function J for f with respect to m_ϕ is

$$J = c\exp(-\phi),$$

where c is the eigenvalue corresponding to m_ϕ, which is proved to be equal to $\exp(P(f,\phi))$ later in Proposition 5.19. We establish in Proposition 5.14 that m_ϕ is a Gibbs state. The measure m_ϕ may not be f-invariant. In Theorem 5.17, we adjust the potential ϕ to get a new potential $\overline{\phi}$ such that there exists an eigenfunction u_ϕ of $\mathscr{L}_{\overline{\phi}}$ with eigenvalue 1. The positive function u_ϕ constructed as the uniform limit of the sequence

$$\left\{\frac{1}{n}\sum_{j=0}^{n-1}\mathscr{L}_{\overline{\phi}}^j(\mathbb{1})\right\}_{n\in\mathbb{N}}$$

is shown to be bounded away from 0 and $+\infty$, and Hölder continuous with the same exponent as that of ϕ. Then we demonstrate that the measure $\mu_\phi = u_\phi m_\phi$ is an f-invariant Gibbs state. Finally, by combining Proposition 5.8 and Proposition 5.19, we prove in Corollary 5.20 that μ_ϕ is an equilibrium state for f and ϕ.

In Sect. 5.3, we establish the uniqueness of the equilibrium state for an expanding Thurston map f and a real-valued Hölder continuous potential ϕ. We use the idea in [PU10] to apply the Gâteaux differentiability of the topological pressure function and some techniques from functional analysis. More precisely, a general fact from functional analysis (recorded in Theorem 5.23) states that for an arbitrary convex continuous function $Q\colon V \to \mathbb{R}$ on a separable Banach space V, there exists a unique continuous linear functional $L\colon V \to \mathbb{R}$ *tangent to* Q *at* $x \in V$ if and only if the function $t \longmapsto Q(x+ty)$ is differentiable at 0 for all y in a subset U of V that is dense in the weak topology on V. One then observes that for each continuous map $g\colon X \to X$ on a compact metric space X, the topological pressure function $P(g,\cdot)\colon C(X) \to \mathbb{R}$ is continuous and convex (see [PU10, Theorem 3.6.1 and Theorem 3.6.2]), and if μ is an equilibrium state for g and $\psi \in C(X)$, then the continuous linear functional $u \longmapsto \int u\,\mathrm{d}\mu$, for $u \in C(X)$, is tangent to $P(g,\cdot)$ at ψ (see [PU10, Proposition 3.6.6]). So in order to verify the uniqueness of the equilibrium state for an expanding Thurston map f and a real-valued Hölder continuous potential ϕ, it suffices to prove that the function $t \longmapsto P(f,\phi+t\gamma)$ is differentiable at 0, for all γ in a suitable subspace of $C(S^2)$. This is established in Theorem 5.35.

Following the procedures in [PU10] to prove Theorem 5.35, we introduce a new potential $\widetilde{\phi}$ induced by ϕ, and establish some uniform bounds in Theorem 5.27 and Lemma 5.30, which are then used to show uniform convergence results in Theorem 5.31 and Lemma 5.33. In some sense, Theorem 5.27 gives a quantitative form of the fact that $\mathscr{L}_{\widetilde{\phi}}$ is *almost periodic* (see Corollary 5.29), and Theorem 5.31 exhibits a uniform version of the contracting behavior of $\mathscr{L}_{\widetilde{\phi}}$ on a codimension-1 subspace of $C(S^2)$. As a by-product, we demonstrate in Corollary 5.32 that for each expanding

Thurston map f and each real-valued Hölder continuous potential ϕ, the operator $\mathcal{L}_{\phi}^{*}$ has a unique eigenmeasure m_{ϕ}. Moreover, the measure μ_{ϕ} is the unique eigenmeasure $m_{\widetilde{\phi}}$ of $\mathcal{L}_{\widetilde{\phi}}^{*}$ with the corresponding eigenvalue 1. Another consequence is Proposition 5.38 which implies (5.3) mentioned earlier.

In Sect. 5.4, we prove that the measure-preserving transformation f of the probability space (S^2, μ_{ϕ}) is exact (Theorem 5.41), where the equilibrium state μ_{ϕ} is non-atomic (Corollary 5.42). It follows in particular that the transformation f is mixing and ergodic (Corollary 5.44). To establish these results, we first show in Proposition 5.39 that

$$m_{\phi}\left(\bigcup_{i=0}^{+\infty} f^{-i}(\mathscr{C})\right) = \mu_{\phi}\left(\bigcup_{i=0}^{+\infty} f^{-i}(\mathscr{C})\right) = 0$$

for each Jordan curve $\mathscr{C} \subseteq S^2$ containing the postcritical points of f that satisfies $f^l(\mathscr{C}) \subseteq \mathscr{C}$ for some $l \in \mathbb{N}$. This proposition is also used in the proof of Theorem 5.45.

Theorem 5.45, the main result of Sect. 5.5, asserts that if ϕ and ψ are two real-valued Hölder continuous functions with the corresponding equilibrium states μ_{ϕ} and μ_{ψ}, respectively, then $\mu_{\phi} = \mu_{\psi}$ if and only if there exists a constant $K \in \mathbb{R}$ such that $\phi - \psi$ and $K\mathbb{1}_{S^2}$ are *co-homologous* in the space $C(S^2)$ of real-valued continuous functions, i.e., $\phi - \psi - K\mathbb{1}_{S^2} = u \circ f - u$ for some $u \in C(S^2)$. For Theorem 5.45, we first formulate a form of the *closing lemma* for expanding Thurston maps (Lemma 5.50). For such maps, we then include in Lemma 5.51 a direct proof of the existence of a point whose forward orbit is dense in S^2. Finally, we give the proof of Theorem 5.45 at the end of the section.

In Sect. 5.6, we first establish in Proposition 5.54 versions of equidistribution of preimages with respect to the equilibrium state, using results we obtain in Sect. 5.3. These results partially generalize Theorem 4.2 where we treat the case for the measure of maximal entropy. At the end of chapter, following the idea of J. Hawkins and M. Taylor [HT03], we prove in Theorem 5.55 that the equilibrium state μ_{ϕ} from Theorem 5.1 is almost surely the limit of

$$\frac{1}{n}\sum_{i=0}^{n-1} \delta_{q_i}$$

as $n \longrightarrow +\infty$ in the weak* topology, where q_0 is an arbitrary fixed point in S^2, and for each $i \in \mathbb{N}_0$, the point q_{i+1} is randomly chosen from the set $f^{-1}(q_i)$ with the probability of each $x \in f^{-1}(q_i)$ being q_{i+1} conditional on q_i proportional to the local degree of f at x times $\exp\big(\phi(x)\big)$. This theorem is an immediate consequence of a theorem of H. Furstenberg and Y. Kifer in [FK83] and the fact that the equilibrium state is the unique Borel probability measure invariant under the adjoint of the Ruelle operator $\mathscr{L}_{\bar{\phi}}$ (Corollary 5.32). A similar result for certain hyperbolic rational maps on the Riemann sphere and the measures of maximal entropy was proved by M. Barnsley [Barn88]. J. Hawkins and M. Taylor generalized it to any rational map on the Riemann sphere of degree $d \geq 2$ [HT03].

5.1 The Assumptions

We state below the hypothesis under which we will develop our theory in most parts of this chapter and Chaps. 6 and 7. We will repeatedly refer to such assumptions in these chapters. We emphasize again that not all assumptions are assumed in all the statements in the subsequent chapters, and that in fact we have to gradually remove the dependence on some of the assumptions before establishing our main results.

The Assumptions

1. $f\colon S^2 \to S^2$ is an expanding Thurston map.
2. $\mathscr{C} \subseteq S^2$ is a Jordan curve containing post f with the property that there exists $n_{\mathscr{C}} \in \mathbb{N}$ such that $f^{n_{\mathscr{C}}}(\mathscr{C}) \subseteq \mathscr{C}$ and $f^m(\mathscr{C}) \nsubseteq \mathscr{C}$ for each $m \in \{1, 2, \ldots, n_{\mathscr{C}} - 1\}$.
3. d is a visual metric on S^2 for f with expansion factor $\Lambda > 1$ and a linear local connectivity constant $L \geq 1$.
4. $\phi \in C^{0,\alpha}(S^2, d)$ is a real-valued Hölder continuous function with an exponent $\alpha \in (0, 1]$.

Observe that by Theorem 2.16, for each f in (1), there exists at least one Jordan curve $\mathscr{C}$ that satisfies (2). Since for a fixed f, the number $n_{\mathscr{C}}$ is uniquely determined by $\mathscr{C}$ in (2), in the remaining part of the paper we will say that a quantity depends on $\mathscr{C}$ even if it also depends on $n_{\mathscr{C}}$.

Recall that the expansion factor Λ of a visual metric d on S^2 for f is uniquely determined by d and f. We will say that a quantity depends on f and d if it depends on Λ.

Note that even though the value of L is not uniquely determined by the metric d, in the remainder of this paper, for each visual metric d on S^2 for f, we will fix a choice of linear local connectivity constant L. We will say that a quantity depends on the visual metric d without mentioning the dependence on L, even though if we had not fixed a choice of L, it would have depended on L as well.

In the discussion below, depending on the conditions we will need, we will sometimes say "Let $f, \mathscr{C}, d, \phi, \alpha$ satisfy the Assumptions.", and sometimes say "Let f and d satisfy the Assumptions.", etc.

5.2 Existence

By the work of P. Haïssinsky and K. Pilgrim [HP09], and M. Bonk and D. Meyer [BM17], we know that there exists a unique measure of maximal entropy μ_f for f, and that

$$h_{\mathrm{top}}(f) = \log(\deg f).$$

In this section, we generalize the existence part of this result to equilibrium states for real-valued Hölder continuous potentials. We prove the uniqueness in the next section.

We first establish the following two distortion lemmas that serve as the cornerstones for all the analysis in the thermodynamical formalism.

Lemma 5.3 *Let f, $\mathcal{C}$, d, L, Λ satisfy the Assumptions. Let $\phi \in C^{0,\alpha}((S^2, d), \mathbb{C})$ be a Hölder continuous function with an exponent $\alpha \in (0, 1]$. Then there exists a constant $C_1 = C_1(f, \mathcal{C}, d, \phi, \alpha)$ depending only on f, $\mathcal{C}$, d, ϕ, and α such that*

$$|S_n\phi(x) - S_n\phi(y)| \leq C_1 d(f^n(x), f^n(y))^\alpha, \tag{5.4}$$

for $n, m \in \mathbb{N}_0$ with $n \leq m$, $X^m \in \mathbf{X}^m(f, \mathcal{C})$, and $x, y \in X^m$. Quantitatively, we choose

$$C_1 = \frac{|\phi|_\alpha C_0}{1 - \Lambda^{-\alpha}}, \tag{5.5}$$

where $C_0 > 1$ is a constant depending only on f, $\mathcal{C}$, and d from Lemma 2.19.

Note that due to the convention described in Sect. 5.1, we do not say that C_1 depends on Λ or $n_{\mathcal{C}}$.

Proof For $n = 0$, inequality (5.4) trivially follows from the definition of S_n.

By Lemma 2.19, we have that for each $m \in \mathbb{N}_0$, each m-tile $X^m \in \mathbf{X}^m(f, \mathcal{C})$, each $x, y \in X^m$, and for $0 \leq j \leq n \leq m$,

$$d(f^j(x), f^j(y)) \leq C_0 \Lambda^{-(n-j)} d(f^n(x), f^n(y)).$$

So $\left|\phi(f^j(x)) - \phi(f^j(y))\right| \leq |\phi|_\alpha C_0^\alpha \Lambda^{-\alpha(n-j)} d(f^n(x), f^n(y))^\alpha$. Thus for each $n \in \mathbb{N}$ with $n \leq m$, we have

$$\begin{aligned}
|S_n\phi(x) - S_n\phi(y)| &\leq \sum_{j=0}^{n-1} \left|\phi(f^j(x)) - \phi(f^j(y))\right| \\
&\leq |\phi|_\alpha C_0^\alpha d(f^n(x), f^n(y))^\alpha \sum_{j=0}^{n-1} \Lambda^{-\alpha(n-j)} \\
&\leq |\phi|_\alpha C_0^\alpha d(f^n(x), f^n(y))^\alpha \sum_{k=0}^{+\infty} \Lambda^{-\alpha k}
\end{aligned}$$

$$\leq \frac{|\phi|_\alpha C_0}{1-\Lambda^{-\alpha}} d(f^n(x), f^n(y))^\alpha$$
$$= C_1 d(f^n(x), f^n(y))^\alpha.$$

□

Lemma 5.4 *Let f, $\mathscr{C}$, d, L, Λ, ϕ, α satisfy the Assumptions. Then there exists $C_2 = C_2(f, \mathscr{C}, d, \phi, \alpha) \geq 1$ depending only on f, $\mathscr{C}$, d, ϕ, and α such that for each $x, y \in S^2$, and each $n \in \mathbb{N}_0$, we have*

$$\frac{\sum\limits_{x' \in f^{-n}(x)} \deg_{f^n}(x') \exp(S_n\phi(x'))}{\sum\limits_{y' \in f^{-n}(y)} \deg_{f^n}(y') \exp(S_n\phi(y'))} \leq \exp\left(4C_1 L d(x,y)^\alpha\right) \leq C_2, \tag{5.6}$$

where C_1 is the constant from Lemma 5.3. Quantitatively, we choose

$$C_2 = \exp\left(4C_1 L \left(\operatorname{diam}_d(S^2)\right)^\alpha\right) = \exp\left(4\frac{|\phi|_\alpha C_0}{1-\Lambda^{-1}} L \left(\operatorname{diam}_d(S^2)\right)^\alpha\right), \tag{5.7}$$

where $C_0 > 1$ is a constant depending only on f, $\mathscr{C}$, and d from Lemma 2.19.

Proof We denote $\Sigma(x,n) = \sum\limits_{x' \in f^{-n}(x)} \deg_{f^n}(x') \exp(S_n\phi(x'))$ for $x \in S^2$ and $n \in \mathbb{N}_0$.

We start with proving the first inequality in (5.6).

Let X^0 be either the black 0-tile X^0_b or the white 0-tile X^0_w in $\mathbf{X}^0(f, \mathscr{C})$. For $n \in \mathbb{N}_0$ and $X^n \in \mathbf{X}^n(f, \mathscr{C})$ with $f^n(X^n) = X^0$, by Proposition 2.6(i), $f^n|_{X^n}$ is a homeomorphism of X^n onto X^0. So for $x, y \in X^0$, there exist unique points $x', y' \in X^n$ with $x' \in f^{-n}(x)$ and $y' \in f^{-n}(y)$. Then by Lemma 5.3, we have

$$\exp\left(S_n\phi(x') - S_n\phi(y')\right) \leq \exp\left(C_1 d(f^n(x'), f^n(y'))^\alpha\right) = \exp\left(C_1 d(x,y)^\alpha\right).$$

Thus $\exp\left(S_n\phi(x')\right) \leq \exp\left(C_1 d(x,y)^\alpha\right) \exp\left(S_n\phi(y')\right)$.

By summing the last inequality over all pairs of x', y' that are contained in the same n-tile X^n with $f^n(X^n) = X^0$, and noting that each x' (resp. y') is contained in exactly $\deg_{f^n}(x')$ (resp. $\deg_{f^n}(y')$) distinct n-tiles X^n with $f^n(X^n) = X^0$, we can conclude that

$$\frac{\Sigma(x,n)}{\Sigma(y,n)} \leq \exp\left(C_1 d(x,y)^\alpha\right).$$

Recall that $f, \mathscr{C}, d, L, \Lambda, \phi, \alpha$ satisfy the Assumptions. We then consider arbitrary $x \in X^0_w$ and $y \in X^0_b$. Since the metric space (S^2, d) is linearly locally connected with a linear local connectivity constant $L \geq 1$, there exists a continuum $E \subseteq S^2$ with $x, y \in E$ and $E \subseteq B_d(x, Ld(x,y))$. We can then fix a point $z \in \mathscr{C} \cap E$. Thus, we have

$$\begin{aligned}\frac{\Sigma(x,n)}{\Sigma(y,n)} &\leq \frac{\Sigma(x,n)}{\Sigma(z,n)}\frac{\Sigma(z,n)}{\Sigma(y,n)} \leq \exp\left(C_1\left(d(x,z)^\alpha + d(z,y)^\alpha\right)\right)\\ &\leq \exp\left(2C_1(\operatorname{diam}_d(E))^\alpha\right) \leq \exp\left(4C_1 L d(x,y)^\alpha\right).\end{aligned}$$

Finally, (5.7) follows from (5.5) in Lemma 5.3. □

Let $f, \mathscr{C}, d, L, \Lambda, \phi, \alpha$ satisfy the Assumptions. We now define the Gibbs states with respect to $f, \mathscr{C}$, and ϕ.

Definition 5.5 A Borel probability measure $\mu \in \mathscr{P}(S^2)$ is a *Gibbs state* with respect to $f, \mathscr{C}$, and ϕ if there exist constants $P_\mu \in \mathbb{R}$ and $C_\mu \geq 1$ such that for each $n \in \mathbb{N}_0$, each n-tile $X^n \in \mathbf{X}^n(f, \mathscr{C})$, and each $x \in X^n$, we have

$$\frac{1}{C_\mu} \leq \frac{\mu(X^n)}{\exp(S_n\phi(x) - nP_\mu)} \leq C_\mu. \tag{5.8}$$

Compare the above definition with the following one, which is used for some classical dynamical systems.

Definition 5.6 A Borel probability measure $\mu \in \mathscr{P}(S^2)$ is a *radial Gibbs state* with respect to f, d, and ϕ if there exist constants $\widetilde{P}_\mu \in \mathbb{R}$ and $\widetilde{C}_\mu \geq 1$ such that for each $n \in \mathbb{N}_0$, and each $x \in S^2$, we have

$$\frac{1}{\widetilde{C}_\mu} \leq \frac{\mu\left(B_d(x, \Lambda^{-n})\right)}{\exp\left(S_n\phi(x) - n\widetilde{P}_\mu\right)} \leq \widetilde{C}_\mu. \tag{5.9}$$

One observes that for each Gibbs state μ with respect to $f, \mathscr{C}$, and ϕ, the constant P_μ is unique. Similarly, the constant $\widetilde{P}_\mu$ is unique for each radial Gibbs state with respect to f, d, and ϕ.

Example 5.7 Let $f : S^2 \to S^2$ be an expanding Thurston map. There exists a unique measure of maximal entropy μ_0 of f (see [HP09, Sect. 3.4 and Sect. 3.5] and [BM17, Theorem 17.11]), which is an equilibrium state for a potential $\phi \equiv 0$. We can show that μ_0 is a Gibbs state for $f, \mathscr{C}, \phi \equiv 0$, whenever $\mathscr{C}$ is a Jordan curve on S^2 containing post f.

Indeed, we know that there exist constants $w, b \in (0, 1)$ depending only on f such that for each $n \in \mathbb{N}_0$, each white n-tile $X^n_w \in \mathbf{X}^n_w(f, \mathscr{C})$, and each black n-tile $X^n_b \in \mathbf{X}^n_b(f, \mathscr{C})$, we have $\mu_0(X^n_w) = w(\deg f)^{-n}$ and $\mu_0(X^n_b) = b(\deg f)^{-n}$ ([BM17, Proposition 17.10 and Theorem 17.11]). Thus μ_0 is a Gibbs state for $f, \mathscr{C}, \phi \equiv 0$, with $P_{\mu_0} = \deg f = h_{\mathrm{top}}(f)$ (see [BM17, Corollary 17.2]).

As we see from the example above, Definition 5.5 is a more appropriate definition for expanding Thurston map. Moreover, we will prove in Proposition 5.21 that the concept of a Gibbs state and that of a radial Gibbs state coincide if and only if f has no periodic critical point.

Proposition 5.8 *Let f, $\mathscr{C}$, $n_{\mathscr{C}}$, d, ϕ, α satisfy the Assumptions. Then for each f-invariant Gibbs state $\mu \in \mathscr{M}(S^2, f)$ with respect to f, $\mathscr{C}$, and ϕ, we have*

$$P_\mu \leq h_\mu(f) + \int \phi \,\mathrm{d}\mu \leq P(f, \phi). \tag{5.10}$$

Proof Note that the second inequality follows from the Variational Principle (3.5) (see for example, [PU10, Theorem 3.4.1] for details).

Let $N = n_{\mathscr{C}}$.

Recall measurable partitions O_n, $n \in \mathbb{N}$, of S^2 defined in (2.8). Since $f^N(\mathscr{C}) \subseteq \mathscr{C}$, it is clear that O_{iN} is a refinement of O_{jN} for $i \geq j \geq 1$. Observe that by Proposition 2.6(i) and induction, we can conclude that for each $k \in \mathbb{N}$,

$$O_N \vee f^{-N}(O_N) \vee \cdots \vee f^{-kN}(O_N) = O_{(k+1)N}. \tag{5.11}$$

So for $m, k \in \mathbb{N}$, the measurable partition $\bigvee\limits_{j=0}^{kN+m-1} f^{-j}(O_N)$ is a refinement of $O_{(k+1)N}$.

By Shannon-McMillan-Breiman Theorem (see for example, [PU10, Theorem 2.5.4]), $h_\mu(f, O_N) = \int f_{\mathscr{I}} \,\mathrm{d}\mu$, where

$$f_{\mathscr{I}} = \lim_{n \to +\infty} \frac{1}{n+1} I\left(\bigvee_{j=0}^{n} f^{-j}(O_N) \right) \quad \mu\text{-a.e. and in } L^1(\mu),$$

and the information function I is defined in (3.2).

Note that for $n \in \mathbb{N}$, $c \in O_n$, and $X^n \in \mathbf{X}^n(f, \mathscr{C})$, either $c \cap X^n = \emptyset$ or $c \subseteq X^n$.

For $n \in \mathbb{N}_0$ and $x \in S^2$, we denote by $X^n(x)$ any one of the n-tiles containing x. Recall that $O_n(x)$ denotes the unique set in the measurable partition O_n that contains x. Note that $O_n(x) \subseteq X^n(x)$. By (5.11) and (5.8) we get

$$\begin{aligned}
\int f_{\mathscr{I}} \,\mathrm{d}\mu &= \lim_{k \to +\infty} \int \frac{1}{kN+1} I\left(\bigvee_{j=0}^{kN} f^{-j}(O_N) \right)(x) \,\mathrm{d}\mu(x) \\
&\geq \liminf_{k \to +\infty} \int \frac{1}{kN+1} I(O_{(k+1)N})(x) \,\mathrm{d}\mu(x) \\
&\geq \liminf_{k \to +\infty} \int \frac{1}{kN+1} \left(-\log \mu \left(X^{(k+1)N}(x) \right) \right) \mathrm{d}\mu(x) \\
&\geq \liminf_{k \to +\infty} \int \frac{(k+1)N P_\mu - S_{(k+1)N}\phi(x) - \log C_\mu}{(k+1)N} \,\mathrm{d}\mu(x) \\
&= P_\mu - \liminf_{k \to +\infty} \frac{1}{(k+1)N} \int S_{(k+1)N}\phi(x) \,\mathrm{d}\mu(x) \\
&= P_\mu - \int \phi \,\mathrm{d}\mu,
\end{aligned}$$

where the last equality comes from (0.3) and the identity $\int \psi \circ f \, d\mu = \int \psi \, d\mu$ for each $\psi \in C(S^2)$ which is equivalent to the fact that μ is f-invariant. Since O_N is a finite measurable partition, the condition that $H_\mu(O_N) < +\infty$ in (3.3) is fulfilled. By (3.3), we get that

$$h_\mu(f) \geq h_\mu(f, O_N) \geq P_\mu - \int \phi \, d\mu.$$

Therefore, $P_\mu \leq h_\mu(f) + \int \phi \, d\mu$. □

Definition 5.9 Let $f : S^2 \to S^2$ be an expanding Thurston map and $\mu \in \mathcal{P}(S^2)$ a Borel probability measure on S^2. A Borel function $J : S^2 \to [0, +\infty)$ is a *Jacobian (function)* for f with respect to μ if for every Borel $A \subseteq S^2$ on which f is injective, the following equation holds:

$$\mu(f(A)) = \int_A J \, d\mu. \tag{5.12}$$

Corollary 5.10 *Let $f : S^2 \to S^2$ be an expanding Thurston map. For each $\psi \in C(S^2)$ and each Borel probability measure $\mu \in \mathcal{P}(S^2)$, if $\mathcal{L}_\psi^*(\mu) = c\mu$ for some constant $c > 0$, then the Jacobian J for f with respect to μ is given by*

$$J(x) = \frac{c}{\deg_f(x) \exp(\psi(x))} \quad \textit{for } x \in S^2. \tag{5.13}$$

Proof We fix some $\mathcal{C}$, d, L, Λ that satisfy the Assumptions.

By Lemma 3.2, for every Borel $A \subseteq S^2$ on which f is injective, we have that $f(A)$ is Borel, and

$$\mu(A) = \frac{\mathcal{L}_\psi^*(\mu)(A)}{c} = \int_{f(A)} \frac{1}{J \circ (f|_A)^{-1}} \, d\mu, \tag{5.14}$$

for the function J given in (5.13).

Since f is injective on each 1-tile $X^1 \in \mathbf{X}^1(f, \mathcal{C})$, and both X^1 and $f(X^1)$ are closed subsets of S^2 by Proposition 2.6, in order to verify (5.12), it suffices to assume that $A \subseteq X$ for some 1-tile $X \in \mathbf{X}^1(f, \mathcal{C})$. Denote the restriction of μ on X by μ_X, i.e., μ_X assigns $\mu(B)$ to each Borel subset B of X.

Let $\widetilde{\mu}$ be a function defined on the set of Borel subsets of X in such a way that $\widetilde{\mu}(B) = \mu(f(B))$ for each Borel $B \subseteq X$. It is clear that $\widetilde{\mu}$ is a Borel measure on X. In this notation, we can write (5.14) as

$$\mu_X(A) = \int_A \frac{1}{J|_X} \, d\widetilde{\mu}, \tag{5.15}$$

for each Borel $A \subseteq X$.

By (5.15), we know that μ_X is absolutely continuous with respect to $\widetilde{\mu}$. On the other hand, since J is positive and uniformly bounded away from $+\infty$ on X, we can conclude that $\widetilde{\mu}$ is absolutely continuous with respect to μ_X. Therefore, by the Radon-Nikodym theorem, for each Borel $A \subseteq X$, we get $\mu(f(A)) = \widetilde{\mu}(A) = \int_A J|_X \, \mathrm{d}\mu_X = \int_A J \, \mathrm{d}\mu$. □

Lemma 5.11 *Let $f: S^2 \to S^2$ be an expanding Thurston map, and $\mathscr{C} \subseteq S^2$ be a Jordan curve containing* post f. *Then there exists a constant $M \in \mathbb{N}$ with the following property:*

For each $m \in \mathbb{N}$ with $m \geq M$, each $n \in \mathbb{N}_0$, and each n-tile $X^n \in \mathbf{X}^n(f, \mathscr{C})$, there exist a white $(n+m)$-tile $X_w^{n+m} \in \mathbf{X}_w^{n+m}(f, \mathscr{C})$ and a black $(n+m)$-tile $X_b^{n+m} \in \mathbf{X}_b^{n+m}(f, \mathscr{C})$ such that $X_w^{n+m} \cup X_b^{n+m} \subseteq \operatorname{inte}(X^n)$.

Proof We fix some d, L, Λ that satisfy the Assumptions.

By Lemma 2.13(v), there exists a constant $C \geq 1$ depending only on f, $\mathscr{C}$, and d such that for each $k \in \mathbb{N}_0$, each k-tile $Z^k \in \mathbf{X}^k(f, \mathscr{C})$, there exists a point $q \in Z^k$ such that

$$B_d(q, C^{-1}\Lambda^{-k}) \subseteq Z^k \subseteq B_d(q, C\Lambda^{-k}).$$

We set $M = \lceil \log_\Lambda(4C^2) \rceil + 1$. We fix an arbitrary $n \in \mathbb{N}$ and an n-tile $X^n \in \mathbf{X}^n(f, \mathscr{C})$. Choose a point $p \in X^n$ with $B_d(p, C^{-1}\Lambda^{-n}) \subseteq X^n \subseteq B_d(p, C\Lambda^{-n})$. Then for each $m \in \mathbb{N}$ with $m \geq M$, we have $4C\Lambda^{-(n+m)} < C^{-1}\Lambda^{-n}$, and we can choose $X^{n+m}, Y^{n+m} \in \mathbf{X}^{n+m}(f, \mathscr{C})$ in such a way that X^{n+m} is the $(n+m)$-tile containing p and $Y^{n+m} \cap X^{n+m} = e^{n+m} \in \mathbf{E}^{n+m}(f, \mathscr{C})$ for each $m > M$. Thus $\operatorname{diam}_d(X^{n+m}) \leq 2C\Lambda^{-(n+m)}$, $\operatorname{diam}_d(Y^{n+m}) \leq 2C\Lambda^{-(n+m)}$, and

$$X^{n+m} \cup Y^{n+m} \subseteq \overline{B_d}\left(p, 4C\Lambda^{-(n+m)}\right) \subseteq B_d\left(p, C^{-1}\Lambda^{-n}\right) \subseteq \operatorname{inte}(X^n).$$

Moreover, exactly one of X^{n+m} and Y^{n+m} is a white $(n+m)$-tile and the other one is a black $(n+m)$-tile. □

Theorem 5.12 *Let $f: S^2 \to S^2$ be an expanding Thurston map, and d be a visual metric on S^2 for f. Let $\phi \in C^{0,\alpha}(S^2, d)$ be a real-valued Hölder continuous function with an exponent $\alpha \in (0, 1]$. Then there exists a Borel probability measure $m_\phi \in \mathscr{P}(S^2)$ such that*

$$\mathscr{L}_\phi^*(m_\phi) = c m_\phi, \tag{5.16}$$

where $c = \langle \mathscr{L}_\phi^*(m_\phi), \mathbb{1} \rangle$. Moreover, any $m_\phi \in \mathscr{P}(S^2)$ that satisfies (5.16) for some $c > 0$ has the following properties:

(i) *The Jacobian for f with respect to m_ϕ is*

$$J(x) = c \exp(-\phi(x)).$$

(ii) $m_\phi\left(\bigcup_{j=0}^{+\infty} f^{-j}(\operatorname{post} f)\right) = 0.$

(iii) *The map f with respect to m_ϕ is forward quasi-invariant (i.e., for each Borel set $A \subseteq S^2$, if $m_\phi(A) = 0$, then $m_\phi(f(A)) = 0$), and nonsingular (i.e., for each Borel set $A \subseteq S^2$, $m_\phi(A) = 0$ if and only if $m_\phi(f^{-1}(A)) = 0$).*

We will see later in Corollary 5.32 that $m_\phi \in \mathcal{P}(S^2)$ satisfying (5.16) is unique. We will also prove in Corollary 5.42 that m_ϕ is non-atomic.

Proof We fix a Jordan curve $\mathcal{C} \subseteq S^2$ that satisfies the Assumptions (see Theorem 2.16 for the existence of such $\mathcal{C}$).

Define $\tau\colon \mathcal{P}(S^2) \to \mathcal{P}(S^2)$ by $\tau(\mu) = \frac{\mathcal{L}_\phi^*(\mu)}{\langle \mathcal{L}_\phi^*(\mu), \mathbb{1}\rangle}$. Then τ is a continuous transformation on the nonempty, convex, compact (in the weak* topology, by Alaoglu's theorem) space $\mathcal{P}(S^2)$ of Borel probability measures on S^2. By the Schauder-Tikhonov Fixed Point Theorem (see for example, [PU10, Theorem 3.1.7]), there exists a measure $m_\phi \in \mathcal{P}(S^2)$ such that $\tau(m_\phi) = m_\phi$. Thus $\mathcal{L}_\phi^*(m_\phi) = c m_\phi$ with $c = \langle \mathcal{L}_\phi^*(m_\phi), \mathbb{1}\rangle$.

By Corollary 5.10, the formula for the Jacobian for f with respect to m_ϕ is

$$J(x) = c(\deg_f(x)\exp(\phi(x)))^{-1}, \qquad \text{for } x \in S^2. \tag{5.17}$$

Since $\bigcup_{j=0}^{+\infty} f^{-j}(\operatorname{post} f)$ is a countable set, the property (ii) follows if we can prove that $m_\phi(\{y\}) = 0$ for each $y \in \bigcup_{j=0}^{+\infty} f^{-j}(\operatorname{post} f)$. Since for each $x \in S^2$,

$$m_\phi(\{f(x)\}) = \frac{c}{\deg_f(x)\exp(\phi(x))} m_\phi(\{x\}), \tag{5.18}$$

it suffices to prove that $m_\phi(\{x\}) = 0$ for each periodic $x \in \operatorname{post} f$.

Suppose that there exists $x \in \operatorname{post} f$ such that $f^l(x) = x$ for some $l \in \mathbb{N}$ and $m_\phi(\{x\}) \neq 0$. Then by (5.18), (2.2), and induction,

$$m_\phi(\{x\}) = \frac{c^l}{\deg_{f^l}(x)\exp(S_l\phi(x))} m_\phi(\{x\}), \tag{5.19}$$

where $S_l\phi$ is defined in (0.3). Thus $c^l = \deg_{f^l}(x)\exp(S_l\phi(x))$.

Similarly, for each $k \in \mathbb{N}$ and each $y \in f^{-kl}(x)$, we have

$$m_\phi(\{x\}) = \frac{c^{kl}}{\deg_{f^{kl}}(y)\exp(S_{kl}\phi(y))} m_\phi(\{y\}). \tag{5.20}$$

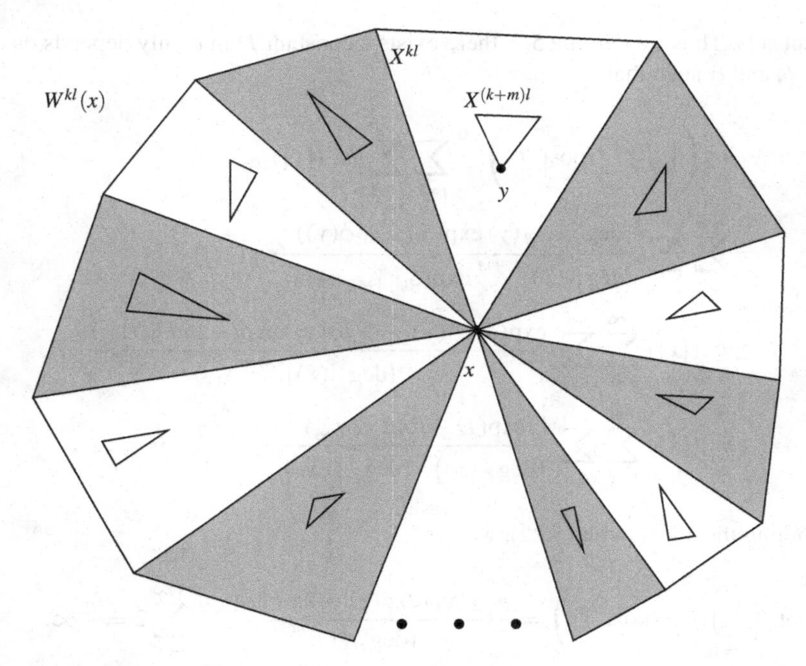

Fig. 5.1 A (kl)-flower $W^{kl}(x)$, with card(post f) = 3

Thus

$$m_\phi(\{y\}) = \frac{\deg_{f^{kl}}(y)\exp(S_{kl}\phi(y))}{\left(\deg_{f^l}(x)\right)^k \exp(S_{kl}\phi(x))} m_\phi(\{x\}). \tag{5.21}$$

Note that for each $k \in \mathbb{N}$, we have $x \in \mathbf{V}^{kl}(f, \mathscr{C})$. The closure of the (kl)-flower $W^{kl}(x)$ of x contains exactly $2\left(\deg_{f^l}(x)\right)^k$ distinct (kl)-tiles whose intersection is $\{x\}$ (see [BM17, Lemma 5.28(i)]). By Lemma 5.11, there exists $m \in \mathbb{N}$ that only depends on f, $\mathscr{C}$, and d such that for each $k \in \mathbb{N}$, each (kl)-tile $X^{kl} \in \mathbf{X}^{kl}(f, \mathscr{C})$ contained in $\overline{W}^{kl}(x)$, there exists a $((k+m)l)$-tile $X^{(k+m)l} \in \mathbf{X}^{(k+m)l}(f, \mathscr{C})$ such that $X^{(k+m)l} \subseteq \operatorname{inte}\left(X^{kl}\right)$. So there exists a unique $y \in X^{(k+m)l} \subseteq \operatorname{inte}\left(X^{kl}\right)$ such that $f^{(k+m)l}(y) = x$ by Proposition 2.6(i), see Fig. 5.1. For each $k \in \mathbb{N}$, we denote by T_k the set consisting of one such y from each (kl)-tile $X^{kl} \subseteq \overline{W}^{kl}(x)$. Note that

$$T_k = 2\left(\deg_{f^l}(x)\right)^k. \tag{5.22}$$

Then $\{T_k\}_{k\in\mathbb{N}}$ is a sequence of subsets of $\bigcup\limits_{j=0}^{+\infty} f^{-j}(\text{post } f)$. Since f is expanding, we can choose an increasing sequence $\{k_i\}_{i\in\mathbb{N}}$ of integers recursively in such a way that $W^{lk_{i+1}}(x) \cap \left(\bigcup\limits_{j=1}^{i} T_{k_j}\right) = \emptyset$ for each $i \in \mathbb{N}$. Then $\{T_{k_i}\}_{i\in\mathbb{N}}$ is a sequence of mutually disjoint sets. Thus by Lemma 5.3, there exists a constant D that only depends on f, $\mathscr{C}$, d, ϕ, and α such that

$$
\begin{aligned}
m_\phi\left(\bigcup_{j=0}^{+\infty} f^{-j}(\operatorname{post} f)\right) &\geq \sum_{i=1}^{+\infty} \sum_{y\in T_{k_i}} m_\phi(\{y\}) \\
&= \sum_{i=1}^{+\infty} \sum_{y\in T_{k_i}} \frac{\deg_{f^{(k_i+m)l}}(y) \exp(S_{(k_i+m)l}\phi(y))}{\left(\deg_{f^l}(x)\right)^{k_i+m} \exp(S_{(k_i+m)l}\phi(x))} m_\phi(\{x\}) \\
&\geq m_\phi(\{x\}) \sum_{i=1}^{+\infty} \sum_{y\in T_{k_i}} \frac{\exp(S_{k_i l}\phi(y) - S_{k_i l}\phi(x)) \exp(-2ml \left\|\phi\right\|_\infty)}{\left(\deg_{f^l}(x)\right)^{k_i+m}} \\
&\geq m_\phi(\{x\}) \sum_{i=1}^{+\infty} \sum_{y\in T_{k_i}} \frac{\exp(D - 2ml \left\|\phi\right\|_\infty)}{\left(\deg_{f^l}(x)\right)^{m} \left(\deg_{f^l}(x)\right)^{k_i}}.
\end{aligned}
$$

Combining the above with (5.22), we get

$$
m_\phi\left(\bigcup_{j=0}^{+\infty} f^{-j}(\operatorname{post} f)\right) = \frac{m_\phi(\{x\}) \exp(D - 2ml \left\|\phi\right\|_\infty)}{\left(\deg_{f^l}(x)\right)^{m}} \sum_{i=1}^{+\infty} 2 = +\infty.
$$

This contradicts the fact that m_ϕ is a finite Borel measure.

Next, in order to prove the formula for the Jacobian for f with respect to m_ϕ in property (i), we observe that by Lemma 3.2 and (5.17), for every Borel set $A \subseteq S^2$ on which f is injective, we have that $f(A)$ is a Borel set and

$$
\begin{aligned}
m_\phi(f(A)) &= m_\phi(f(A) \setminus \operatorname{post} f) = m_\phi(f(A \setminus (\operatorname{post} f \cup \operatorname{crit} f))) \\
&= \int_{A\setminus(\operatorname{post} f \cup \operatorname{crit} f)} c \exp(-\phi) \, \mathrm{d}m_\phi = \int_A c \exp(-\phi) \, \mathrm{d}m_\phi.
\end{aligned}
$$

Finally, we prove the last property. Fix a Borel set $A \subseteq S^2$ with $m_\phi(A) = 0$. For each 1-tile $X^1 \in \mathbf{X}^1(f, \mathscr{C})$, the map f is injective both on $A \cap X^1$ and on $f^{-1}(A) \cap X^1$ by Proposition 2.6(i). So it follows from the formula for the Jacobian that $m_\phi\left(f\left(A \cap X^1\right)\right) = 0$ and $m_\phi\left(f^{-1}(A) \cap X^1\right) = 0$. Thus $m_\phi(f(A)) = 0$ and $\phi(f^{-1}(A)) = 0$. It is clear now that f is forward quasi-invariant and nonsingular with respect to m_ϕ. □

Proposition 5.13 *Let f, d, ϕ, α satisfy the Assumptions. Let m_ϕ be a Borel probability measure defined in Theorem 5.12 with $\mathscr{L}_\phi^*(m_\phi) = c m_\phi$ where $c = \langle \mathscr{L}_\phi^*(m_\phi), \mathbb{1} \rangle$. Then for every Borel set $A \subseteq S^2$, we have*

$$
\frac{1}{\deg f} \int_A J \, \mathrm{d}m_\phi \leq m_\phi(f(A)) \leq \int_A J \, \mathrm{d}m_\phi.
$$

where $J = c\exp(-\phi)$.

Proof We fix a Jordan curve $\mathscr{C} \subseteq S^2$ that satisfies the Assumptions.

The second inequality follows from Definition 5.9 and Theorem 5.12.

Let $B = f(A) \cap X_w^0$ and $C = f(A) \cap \operatorname{inte}(X_b^0)$, where $X_w^0, X_b^0 \in \mathbf{X}^0(f, \mathscr{C})$ are the white 0-tile and the black 0-tile, respectively. Then $B \cap C = \emptyset$ and $B \cup C = f(A)$. For each white 1-tile $X_w^1 \in \mathbf{X}_w^1(f, \mathscr{C})$ and each black 1-tile $X_b^1 \in \mathbf{X}_b^1(f, \mathscr{C})$, we have

$$\int_{f^{-1}(B)\cap X_w^1} J \,\mathrm{d}m_\phi = m_\phi(B), \qquad \int_{f^{-1}(C)\cap \operatorname{inte}(X_b^1)} J \,\mathrm{d}m_\phi = m_\phi(C),$$

by Definition 5.9 and Theorem 5.12. Then the first inequality follows from the fact that $\operatorname{card}\left(\mathbf{X}_w^1(f, \mathscr{C})\right) = \operatorname{card}\left(\mathbf{X}_b^1(f, \mathscr{C})\right) = \deg f$ (see (2.4)). □

Proposition 5.14 *Let f, $\mathscr{C}$, d, ϕ, α satisfy the Assumptions. Let m_ϕ be a Borel probability measure defined in Theorem 5.12 which satisfies $\mathcal{L}_\phi^*(m_\phi) = cm_\phi$ where $c = \langle \mathcal{L}_\phi^*(m_\phi), \mathbb{1}\rangle$. Then m_ϕ is a Gibbs state with respect to f, $\mathscr{C}$, and ϕ, with*

$$P_{m_\phi} = \log c = \lim_{n\to+\infty} \frac{1}{n} \log \mathcal{L}_\phi^n(\mathbb{1})(y), \tag{5.23}$$

for each $y \in S^2$.

In particular, since the existence of m_ϕ in Theorem 5.12 is independent of $\mathscr{C}$, this proposition asserts that m_ϕ is a Gibbs state with respect to f, $\mathscr{C}$, and ϕ, for each $\mathscr{C}$ that satisfies the Assumptions. In general, it is not clear that a Gibbs state with respect to f, $\mathscr{C}_1$, and ϕ is also a Gibbs state with respect to f, $\mathscr{C}_2$, and ϕ, even though the answer is positive in the case when f has no periodic critical points as shown in Corollary 5.22.

Proof We first need to prove that $\mu = m_\phi$ satisfies (5.8).

We observe that

$$m_\phi(f^i(B)) = \int_B \exp(i\log c - S_i\phi(x)) \,\mathrm{d}m_\phi(x) \tag{5.24}$$

for $n \in \mathbb{N}$, $i \in \{0, 1, \dots, n\}$, and each Borel set $B \subseteq S^2$ on which f^n is injective. Indeed, by the formula for the Jacobian in Theorem 5.12, for a given Borel set $A \subseteq S^2$ on which f is injective, we have

$$\int_{f(A)} g(x) \,\mathrm{d}m_\phi(x) = \int_A (g\circ f)(x) \exp(\log c - \phi(x)) \,\mathrm{d}m_\phi(x)$$

for each simple function g on S^2, thus also for each integrable function g. We establish (5.24) for each $n \in \mathbb{N}$ and each Borel set $B \subseteq S^2$ on which f^n is injective by induction on i. For $i = 0$, equation (5.24) holds trivially. Assume that (5.24) is established for some $i \in \{0, 1, \dots, n-1\}$, then since f^i is injective on $f(B)$, we get

$$\begin{aligned} m_\phi(f^{i+1}(B)) &= \int_{f(B)} \exp(i \log c - S_i\phi(x)) \, \mathrm{d}m_\phi(x) \\ &= \int_B \exp((i+1) \log c - S_{i+1}\phi(x)) \, \mathrm{d}m_\phi(x). \end{aligned}$$

The induction is now complete. In particular, by Proposition 2.6(i),

$$m_\phi(f^n(X^n)) = \int_{X^n} \exp(n \log c - S_n\phi(x)) \, \mathrm{d}m_\phi(x),$$

for $n \in \mathbb{N}$ and $X^n \in \mathbf{X}^n(f, \mathscr{C})$.

Thus by Lemma 5.3, there exists a constant $C \geq 1$ such that for each $n \in \mathbb{N}_0$, each $X^n \in \mathbf{X}^n(f, \mathscr{C})$, and each $x \in X^n$,

$$m_\phi(f^n(X^n)) \geq C^{-1} \exp(n \log c - S_n\phi(x)) m_\phi(X^n)$$

and

$$m_\phi(f^n(X^n)) \leq C \exp(n \log c - S_n\phi(x)) m_\phi(X^n).$$

Note that $f^n(X^n)$ is either the black 0-tile $X_b^0 \in \mathbf{X}^0(f, \mathscr{C})$ or the white 0-tile $X_w^0 \in \mathbf{X}^0(f, \mathscr{C})$. Both X_b^0 and X_w^0 are of positive m_ϕ-measure, for otherwise, suppose that $m_\phi(X^0) = 0$ for some $X^0 \in \mathbf{X}^0(f, \mathscr{C})$, then by Proposition 5.13, $m_\phi(f^j(X^0)) = 0$, for each $j \in \mathbb{N}$. Then by Lemma 5.11, $m_\phi(S^2) = 0$, a contradiction. Hence (5.8) follows, and m_ϕ is a Gibbs state with respect to f, $\mathscr{C}$, and ϕ, with $P_{m_\phi} = \log c$.

To finish the proof, we note that by (3.8) and Lemma 5.4, for each $x, y \in S^2$ and each $n \in \mathbb{N}_0$, we have

$$\frac{1}{C_2} \leq \frac{\mathcal{L}_\phi^n(\mathbb{1})(x)}{\mathcal{L}_\phi^n(\mathbb{1})(y)} \leq C_2, \tag{5.25}$$

where C_2 is a constant depending only on f, $\mathscr{C}$, d, ϕ, and α from Lemma 5.4. Since $\langle m_\phi, \mathcal{L}_\phi^n(\mathbb{1})\rangle = \langle (\mathcal{L}_\phi^*)^n(m_\phi), \mathbb{1}\rangle = \langle c^n m_\phi, \mathbb{1}\rangle = c^n$, by (3.8) and (5.25), we have that for each arbitrarily chosen $y \in S^2$,

$$\begin{aligned} \log c &= \lim_{n \to +\infty} \frac{1}{n} \log \int \mathcal{L}_\phi^n(\mathbb{1})(x) \, \mathrm{d}m_\phi(x) \\ &= \lim_{n \to +\infty} \frac{1}{n} \log \int \mathcal{L}_\phi^n(\mathbb{1})(y) \, \mathrm{d}m_\phi(x) \\ &= \lim_{n \to +\infty} \frac{1}{n} \log \mathcal{L}_\phi^n(\mathbb{1})(y). \end{aligned} \tag{5.26}$$

□

Corollary 5.15 *Let f, d, ϕ, α satisfy the Assumptions. Then the limit*

$$\lim_{n\to+\infty} \frac{1}{n} \log \mathcal{L}_{\phi}^{n}(\mathbb{1})(x)$$

exists for each $x \in S^2$ and is independent of $x \in S^2$.

We denote the limit as $D_\phi \in \mathbb{R}$.

Proof By Theorem 5.12, there exists a measure m_ϕ such as the one in Proposition 5.14. The limit then clearly only depends on f, d, ϕ, and α, and in particular, does not depend on $\mathscr{C}$ or the choice of m_ϕ. □

Let $f, \mathscr{C}, d, \phi, \alpha$ satisfy the Assumptions. We define the function

$$\overline{\phi} = \phi - D_\phi \in C^{0,\alpha}(S^2, d). \tag{5.27}$$

Then

$$\mathcal{L}_{\overline{\phi}} = e^{-D_\phi} \mathcal{L}_\phi. \tag{5.28}$$

If m_ϕ is a Gibbs state from Theorem 5.12, then by Proposition 5.14 and Corollary 5.15 we have

$$\mathcal{L}_{\phi}^{*}(m_\phi) = e^{D_\phi} m_\phi = e^{P_{m_\phi}} m_\phi, \tag{5.29}$$

and

$$\mathcal{L}_{\overline{\phi}}^{*}(m_\phi) = m_\phi, \tag{5.30}$$

since for each $u \in C(S^2)$,

$$\langle \mathcal{L}_{\overline{\phi}}^{*}(m_\phi), u\rangle = \langle m_\phi, \mathcal{L}_{\overline{\phi}}(u)\rangle = e^{-D_\phi}\langle m_\phi, \mathcal{L}_\phi(u)\rangle = e^{-D_\phi}\langle \mathcal{L}_{\phi}^{*}(m_\phi), u\rangle = \langle m_\phi, u\rangle.$$

We summarize in the following lemma the properties of $\mathcal{L}_{\overline{\phi}}$ that we will need.

Lemma 5.16 *Let f, $\mathscr{C}$, d, L, Λ, ϕ, α satisfy the Assumptions. Then there exists a constant $C_3 = C_3(f, \mathscr{C}, d, \phi, \alpha)$ depending only on f, $\mathscr{C}$, d, ϕ, and α such that for each $x, y \in S^2$ and each $n \in \mathbb{N}_0$ the following equations are satisfied*

$$\frac{\mathcal{L}_{\overline{\phi}}^{n}(\mathbb{1})(x)}{\mathcal{L}_{\overline{\phi}}^{n}(\mathbb{1})(y)} \leq \exp\left(4C_1 L d(x,y)^\alpha\right) \leq C_2, \tag{5.31}$$

$$\frac{1}{C_2} \leq \mathcal{L}_{\overline{\phi}}^{n}(\mathbb{1})(x) \leq C_2, \tag{5.32}$$

$$\left|\mathcal{L}_{\overline{\phi}}^{n}(\mathbb{1})(x) - \mathcal{L}_{\overline{\phi}}^{n}(\mathbb{1})(y)\right| \leq C_2\left(\exp\left(4C_1 L d(x,y)^\alpha\right) - 1\right) \leq C_3 d(x,y)^\alpha, \tag{5.33}$$

where C_1, C_2 are constants in Lemma 5.3 and Lemma 5.4 depending only on f, $\mathscr{C}$, d, ϕ, and α.

Proof Inequality (5.31) follows from (5.28), (3.8), and Lemma 5.4.

To prove (5.32), we choose a Gibbs state m_ϕ with respect to f, $\mathscr{C}$, and ϕ from Theorem 5.12. Then by (5.30) and (5.31), we have

$$\mathscr{L}_{\overline{\phi}}^n(\mathbb{1})(x) \leq C_2\big\langle m_\phi, \mathscr{L}_{\overline{\phi}}^n(\mathbb{1})\big\rangle = C_2\big\langle(\mathscr{L}_{\overline{\phi}}^*)^n(m_\phi), \mathbb{1}\big\rangle = C_2\langle m_\phi, \mathbb{1}\rangle = C_2.$$

The first inequality in (5.32) can be proved similarly.

Applying (5.31) and (5.32), we get

$$\begin{aligned}\mathscr{L}_{\overline{\phi}}^n(\mathbb{1})(x) - \mathscr{L}_{\overline{\phi}}^n(\mathbb{1})(y) &= \left(\frac{\mathscr{L}_{\overline{\phi}}^n(\mathbb{1})(x)}{\mathscr{L}_{\overline{\phi}}^n(\mathbb{1})(y)} - 1\right)\mathscr{L}_{\overline{\phi}}^n(\mathbb{1})(x)\\ &\leq C_2\left(\exp\left(4C_1 L d(x,y)^\alpha\right) - 1\right)\\ &\leq C_3 d(x,y)^\alpha,\end{aligned}$$

for some constant C_3 depending only on L, C_1, C_2, and $\operatorname{diam}_d(S^2)$. □

We can now prove the existence of an f-invariant Gibbs state.

Theorem 5.17 *Let $f\colon S^2 \to S^2$ be an expanding Thurston map and $\mathscr{C} \subseteq S^2$ be a Jordan curve containing* post f *with the property that $f^{n_{\mathscr{C}}}(\mathscr{C}) \subseteq \mathscr{C}$ for some $n_{\mathscr{C}} \in \mathbb{N}$ and $f^i(\mathscr{C}) \nsubseteq \mathscr{C}$ for each $i \in \{1, 2, \dots, n_{\mathscr{C}} - 1\}$. Let d be a visual metric on S^2 for f with expansion factor $\Lambda > 1$. Let $\phi \in C^{0,\alpha}(S^2, d)$ be a real-valued Hölder continuous function with an exponent $\alpha \in (0, 1]$. Then the sequence $\left\{\frac{1}{n}\sum\limits_{j=0}^{n-1} \mathscr{L}_{\overline{\phi}}^j(\mathbb{1})\right\}_{n\in\mathbb{N}}$ converges uniformly to a function $u_\phi \in C^{0,\alpha}(S^2, d)$, which satisfies*

$$\mathscr{L}_{\overline{\phi}}(u_\phi) = u_\phi, \tag{5.34}$$

and

$$\frac{1}{C_2} \leq u_\phi(x) \leq C_2, \qquad \text{for each } x \in S^2, \tag{5.35}$$

where $C_2 \geq 1$ is a constant from Lemma 5.4. Moreover, if we let m_ϕ be a Gibbs state from Theorem 5.12, then

$$\int u_\phi \,\mathrm{d}m_\phi = 1, \tag{5.36}$$

and $\mu_\phi = u_\phi m_\phi$ is a Gibbs state with respect to f, $\mathscr{C}$, and ϕ, with

$$P_{\mu_\phi} = P_{m_\phi} = D_\phi = \lim_{n\to+\infty} \frac{1}{n} \log \mathscr{L}_\phi^n(\mathbb{1})(y), \tag{5.37}$$

for each $y \in S^2$, and

$$f_*(\mu_\phi) = \mu_\phi. \tag{5.38}$$

Proof In order to prove this theorem, we first establish (5.34), (5.35), and (5.36) for a subsequential limit of the sequence $\left\{ \frac{1}{n} \sum_{j=0}^{n-1} \mathcal{L}_{\overline{\phi}}^j(\mathbb{1}) \right\}_{n \in \mathbb{N}}$, then prove the above sequence has a unique subsequential limit, and finally justify (5.37) and (5.38).

Define, for each $n \in \mathbb{N}$, $u_n = \frac{1}{n} \sum_{j=0}^{n-1} \mathcal{L}_{\overline{\phi}}^j(\mathbb{1})$. Then $\{u_n\}_{n \in \mathbb{N}}$ is a uniformly bounded sequence of equicontinuous functions on S^2 by (5.32) and (5.33). By the Arzelà-Ascoli Theorem, there exists a continuous function $u_\phi \in C(S^2)$ and an increasing sequence $\{n_i\}_{i \in \mathbb{N}}$ in $\mathbb{N}$ such that $u_{n_i} \to u_\phi$ uniformly on S^2 as $i \longrightarrow +\infty$.

To prove (5.34), we note that by the definition of u_n and (5.32), we have that for each $i \in \mathbb{N}$,

$$\left\| \mathcal{L}_{\overline{\phi}}(u_{n_i}) - u_{n_i} \right\|_\infty = \frac{1}{n_i} \left\| \mathcal{L}_{\overline{\phi}}^{n_i}(\mathbb{1}) - \mathbb{1} \right\|_\infty \leq \frac{1 + C_2}{n_i}.$$

By letting $i \longrightarrow +\infty$, we can conclude that $\left\| \mathcal{L}_{\overline{\phi}}(u_\phi) - u_\phi \right\|_\infty = 0$. Thus (5.34) holds.

By (5.32), we have that $C_2^{-1} \leq u_n(x) \leq C_2$, for each $n \in \mathbb{N}$ and each $x \in S^2$. Thus (5.35) follows.

By (5.30) and definition of u_n, we have $\int u_n \, dm_\phi = \int \mathbb{1} \, dm_\phi = 1$ for each $n \in \mathbb{N}$. Then by the Lebesgue Dominated Convergence Theorem, we can conclude that

$$\int u_\phi \, dm_\phi = \lim_{i \to +\infty} \int u_{n_i} \, dm_\phi = 1,$$

proving (5.36).

Next, we prove that u_ϕ is the unique subsequential limit of the sequence $\{u_n\}_{n \in \mathbb{N}}$ with respect to the uniform norm. Suppose that v_ϕ is another subsequential limit of u_n, $n \in \mathbb{N}$, with respect to the uniform norm. Then v_ϕ is also a continuous function on S^2 satisfying (5.34), (5.35), and (5.36) by the argument above. Let

$$t = \sup\{s \in \mathbb{R} \mid u_\phi(x) - s v_\phi(x) > 0 \text{ for all } x \in S^2\}.$$

By (5.35), $t \in (0, +\infty)$. Then there is a point $y \in S^2$ such that $u_\phi(y) - t v_\phi(y) = 0$. By (3.8) and the equation

$$\mathcal{L}_{\overline{\phi}}(u_\phi - t v_\phi) = u_\phi - t v_\phi,$$

which comes from (5.34), we get that $u_\phi(z) - t v_\phi(z) = 0$ for all $z \in f^{-1}(y)$. Inductively, we can conclude that $u_\phi(z) - t v_\phi(z) = 0$ for all $z \in \bigcup_{i \in \mathbb{N}} f^{-i}(y)$. By

Lemma 2.12, the set $\bigcup_{i \in \mathbb{N}} f^{-i}(y)$ is dense in S^2. Hence $u_\phi = t v_\phi$ on S^2. Since both u_ϕ and v_ϕ satisfy (5.36), we get $t = 1$. Thus $u_\phi = v_\phi$. We have proved that u_n converges to u_ϕ uniformly as $n \longrightarrow +\infty$.

We now prove that $u_\phi \in C^{0,\alpha}(S^2, d)$. Indeed, for each $\varepsilon > 0$, there exists $n \in \mathbb{N}$ such that $\|u_n - u_\phi\|_\infty < \varepsilon$. Then by (5.33), for each $x, y \in S^2$, we have

$$\begin{aligned} |u_\phi(x) - u_\phi(y)| &\leq |u_\phi(x) - u_n(x)| + |u_n(x) - u_n(y)| + |u_n(y) - u_\phi(y)| \\ &\leq 2\varepsilon + \frac{1}{n} \sum_{j=0}^{n-1} \left| \mathcal{L}_{\overline{\phi}}^j(\mathbb{1})(x) - \mathcal{L}_{\overline{\phi}}^j(\mathbb{1})(y) \right| \leq 2\varepsilon + C_3 d(x, y)^\alpha, \end{aligned}$$

where C_3 is a constant in (5.33) from Lemma 5.16. By letting $\varepsilon \longrightarrow 0$, we conclude that $u_\phi \in C^{0,\alpha}(S^2, d)$.

Since m_ϕ is a Gibbs state by Proposition 5.14, then by (5.35), $\mu_\phi = u_\phi m_\phi$ is also a Gibbs state with $P_{\mu_\phi} = P_{m_\phi} = D_\phi = \lim_{n \to +\infty} \frac{1}{n} \log \mathcal{L}_\phi^n(\mathbb{1})(y)$ for each $y \in S^2$, proving (5.37).

Finally we need to prove that μ_ϕ is f-invariant. It suffices to prove that $\langle \mu_\phi, g \circ f \rangle = \langle \mu_\phi, g \rangle$ for each $g \in C(S^2)$. Indeed, by (5.30), (5.34), and (3.9), we get

$$\begin{aligned} \langle \mu_\phi, g \circ f \rangle &= \langle m_\phi, u_\phi (g \circ f) \rangle = \left\langle \mathcal{L}_{\overline{\phi}}^*(m_\phi), u_\phi (g \circ f) \right\rangle \\ &= \left\langle m_\phi, \mathcal{L}_{\overline{\phi}}(u_\phi (g \circ f)) \right\rangle = \left\langle m_\phi, g \mathcal{L}_{\overline{\phi}}(u_\phi) \right\rangle = \langle m_\phi, g u_\phi \rangle = \langle \mu_\phi, g \rangle. \end{aligned}$$

□

Remark 5.18 By a similar argument to that in the proof of the uniqueness of the subsequential limit of $\left\{ \frac{1}{n} \sum_{j=0}^{n-1} \mathcal{L}_{\overline{\phi}}^j(\mathbb{1}) \right\}_{n \in \mathbb{N}}$, one can show that u_ϕ is the unique eigenfunction, upto scalar multiplication, of $\mathcal{L}_{\overline{\phi}}$ corresponding to the eigenvalue 1.

We now get the following characterization of the topological pressure $P(f, \phi)$ of an expanding Thurston map f with respect to a Hölder continuous potential ϕ.

Proposition 5.19 *Let f, d, ϕ, α satisfy the Assumptions. Then for each $x \in S^2$, we have*

$$P(f, \phi) = \lim_{n \to +\infty} \frac{1}{n} \log \sum_{y \in f^{-n}(x)} \deg_{f^n}(y) \exp(S_n \phi(y)) = D_\phi. \tag{5.39}$$

Recall that $D_\phi = P_{m_\phi} = P_{\mu_\phi} = \log c = \log \int \mathcal{L}_\phi(\mathbb{1}) \, \mathrm{d}m_\phi$, using the notation from Proposition 5.14 and Theorem 5.17.

Proof We fix a Jordan curve $\mathscr{C} \subseteq S^2$ that satisfies the Assumptions (see Theorem 2.16 for the existence of such $\mathscr{C}$).

By Corollary 5.15 and (3.8), for each $x \in S^2$, the limit in (5.39) always exists and is equal to D_ϕ, independent of x. Moreover, for an f-invariant Gibbs measure μ_ϕ from Theorem 5.17 with $P_{\mu_\phi} = D_\phi$, we get from Proposition 5.8 that

$$D_\phi = P_{\mu_\phi} \leq P(f, \phi). \tag{5.40}$$

Now it suffices to prove $D_\phi \geq P(f, \phi)$.

Note that by Lemma 2.13(ii), there is a constant $C \geq 1$ depending only on $f, \mathscr{C}$, and d such that for each $n \in \mathbb{N}_0$ and each n-tile $X^n \in \mathbf{X}^n(f, \mathscr{C})$, we have $C^{-1}\Lambda^{-n} \leq \operatorname{diam}_d(X^n) \leq C\Lambda^{-n}$.

Fix $m \in \mathbb{N}$, let $\varepsilon = C\Lambda^{-m}$. For each $n \in \mathbb{N}_0$, let $F_n(m)$ be a maximal (n, ε)-separated subset of S^2.

We claim that if $y_1, y_2 \in F_n(m)$ and $y_1, y_2 \in X^{m+n}$ for some $(m+n)$-tile X^{m+n} in $\mathbf{X}^{m+n}(f, \mathscr{C})$, then $y_1 = y_2$.

Indeed, for each integer $j \in [0, n-1]$, we have that

$$d(f^j(y_1), f^j(y_2)) \leq \operatorname{diam}_d(f^j(X^{m+n})) \leq C\Lambda^{-(m+n-j)} < \varepsilon. \tag{5.41}$$

So suppose that $y_1 \neq y_2$, then y_1, y_2 would not be (n, ε)-separated, a contradiction.

We fix $x \in \operatorname{inte}(X^0_w)$ and $y \in \operatorname{inte}(X^0_b)$ where X^0_w and X^0_b are the white 0-tile and black 0-tile in $\mathbf{X}^0(f, \mathscr{C})$, respectively. We can now construct an injective map $i_n \colon F_n(m) \to f^{-(m+n)}(x) \cup f^{-(m+n)}(y)$ for each $n \in \mathbb{N}$ by demanding that $z \in F_n(m)$ and $i_n(z) \in f^{-(m+n)}(x) \cup f^{-(m+n)}(y)$ be in the same $(m+n)$-tile. Since for each $X^{m+n} \in \mathbf{X}^{m+n}(f, \mathscr{C})$, $\operatorname{card}\left(X^{m+n} \cap (f^{-(m+n)}(x) \cup f^{-(m+n)}(y))\right) = 1$, it follows that i_n is well-defined (but not necessarily uniquely defined) for each $n \in \mathbb{N}$. Thus by Lemma 5.3 and Lemma 5.4, we have that for each $n \in \mathbb{N}$,

$$\begin{aligned}
\sum_{z \in F_n(m)} \exp(S_n\phi(z)) &\leq C_4 \sum_{z \in f^{-(m+n)}(x) \cup f^{-(m+n)}(y)} \exp(S_n\phi(z)) \\
&\leq C_4 e^{m\|\phi\|_\infty} \left(\sum_{z \in f^{-(m+n)}(x)} \exp(S_{m+n}\phi(z)) + \sum_{z \in f^{-(m+n)}(y)} \exp(S_{m+n}\phi(z)) \right) \\
&\leq C_4(1 + C_2)\exp(m\|\phi\|_\infty) \sum_{z \in f^{-(m+n)}(x)} \exp(S_{m+n}\phi(z)),
\end{aligned}$$

where $C_4 = \exp\left(C_1\left(\operatorname{diam}_d(S^2)\right)^\alpha\right)$, and C_1, C_2 are constants from Lemma 5.3 and Lemma 5.4. By taking logarithm and next dividing by n on both sides, then taking $n \longrightarrow +\infty$ and finally taking $m \longrightarrow +\infty$ to make $\varepsilon \longrightarrow 0$, we get from (3.1) that

$$\begin{aligned}
P(f, \phi) &= \lim_{m \to +\infty} \liminf_{n \to +\infty} \frac{1}{n} \log \sum_{w \in F_n(m)} \exp(S_n\phi(w)) \\
&\leq \limsup_{m \to +\infty} \liminf_{n \to +\infty} \frac{1}{n} \log \sum_{z \in f^{-(m+n)}(x)} \exp(S_{m+n}\phi(z))
\end{aligned}$$

$$= \limsup_{m\to+\infty} \liminf_{n\to+\infty} \frac{1}{m+n} \log \sum_{z\in f^{-(m+n)}(x)} \exp(S_{m+n}\phi(z))$$

$$= \limsup_{m\to+\infty} \liminf_{n\to+\infty} \frac{1}{n} \log \sum_{z\in f^{-n}(x)} \exp(S_n\phi(z))$$

$$= D_\phi,$$

where the last equality follows from Corollary 5.15, (3.8), and the fact that $x \notin \operatorname{post} f$. □

The following corollary gives the existence part of Theorem 5.1.

Corollary 5.20 *Let $f : S^2 \to S^2$ be an expanding Thurston map and d be a visual metric on S^2 for f. Let $\phi \in C^{0,\alpha}(S^2, d)$ be a real-valued Hölder continuous function with an exponent $\alpha \in (0, 1]$. Then there exists an equilibrium state for f and ϕ. In fact, any measure μ_ϕ defined in Theorem 5.17 is an equilibrium state for f and ϕ.*

Proof We fix a Jordan curve $\mathscr{C} \subseteq S^2$ that satisfies the Assumptions (see Theorem 2.16 for the existence of such $\mathscr{C}$). Consider an f-invariant Gibbs state μ_ϕ with respect to f, $\mathscr{C}$, and ϕ from Theorem 5.17. Then by Theorem 5.17 and Proposition 5.19, we have $P_{\mu_\phi} = D_\phi = P(f, \phi)$. Then by Proposition 5.8, we have $P_{\mu_\phi} = h_{\mu_\phi} + \int \phi \,\mathrm{d}\mu_\phi = P(f, \phi)$. Therefore, μ_ϕ is an equilibrium state for f and ϕ. □

We end this section by proving in Proposition 5.21 that the concept of a Gibbs state and that of a radial Gibbs state coincide if and only if the expanding Thurston map has no periodic critical point. The proof of the forward implication relies on the existence of Gibbs states for f, $\mathscr{C}$, and ϕ that satisfy the Assumptions proved in Proposition 5.14.

Proposition 5.21 *Let f, $\mathscr{C}$, d, ϕ, α satisfy the Assumptions. Then a radial Gibbs state μ with respect to f, d, and ϕ must be a Gibbs state with respect to f, $\mathscr{C}$, and ϕ, with $\widetilde{P}_\mu = P_\mu$. Moreover, the following are equivalent:*

1. *f has no periodic critical point.*
2. *A Borel probability measure $\mu \in \mathscr{P}(S^2)$ is a Gibbs state with respect to f, $\mathscr{C}$, and ϕ if and only if it is a radial Gibbs state with respect to f, d, and ϕ.*
3. *There exists a radial Gibbs state for f, d, and ϕ.*

The implication from (1) to (2) generalizes Proposition 18.2 in [BM17], which states that for an expanding Thurston map f with no periodic critical point and with the measure of maximal entropy μ and a visual metric d, the metric measure space (S^2, d, μ) is Ahlfors regular.

Proof By Lemma 2.13(v), there exists a constant $C \geq 1$ such that for each $n \in \mathbb{N}_0$, and each n-tile $X^n \in \mathbf{X}^n$, there exists a point $p \in X^n$ with

$$B_d(p, C^{-1}\Lambda^{-n}) \subseteq X^n \subseteq B_d(p, C\Lambda^{-n}).$$

Thus there exists $m_1 \in \mathbb{N}$ such that for each $n \in \mathbb{N}_0$, each $X^n \in \mathbf{X}^n$, there exists $p \in X^n$ such that

$$B_d\left(p, \Lambda^{-(n+m_1)}\right) \subseteq X^n \subseteq B_d\left(p, \Lambda^{-(n-m_1)}\right). \tag{5.42}$$

On the other hand, by Lemma 2.13(iv), there exists $m_2 \in \mathbb{N}$ such that for each $x \in S^2$ and each $n \in \mathbb{N}_0$, we have

$$U^{n+m_2}(x) \subseteq B_d(x, \Lambda^{-n}) \subseteq U^{n-m_2}(x), \tag{5.43}$$

where the sets $U^l(x)$ for $l \in \mathbb{N}_0$ and $x \in S^2$ are defined in (2.7).

Note that for each $n \in \mathbb{N}_0$ and each $y \in U^n(x)$, by choosing $z \in Y^n \cap X^n$ with $X^n, Y^n \in \mathbf{X}^n$ and $x \in X^n$, $y \in Y^n$, and applying Lemma 5.3, we get

$$|S_n\phi(x) - S_n\phi(y)| \leq |S_n\phi(x) - S_n\phi(z)| + |S_n\phi(z) - S_n\phi(y)| \leq 2C_1\left(\operatorname{diam}_d(S^2)\right)^{\alpha},$$

where C_1 is a constant from Lemma 5.3.

If μ is a radial Gibbs state with constants $\widetilde{P}_\mu$ and $\widetilde{C}_\mu$, then for each $n \in \mathbb{N}_0$ and each n-tile $X^n \in \mathbf{X}^n$, there exists $p \in X^n$ such that

$$\begin{aligned}\mu(X^n) &\leq \mu\left(B_d\left(p, \Lambda^{-(n-m_1)}\right)\right) \leq \widetilde{C}_\mu \exp\left(S_{n-m_1}\phi(x) - (n-m_1)\widetilde{P}_\mu\right) \\ &\leq \widetilde{C}_\mu \exp\left(m_1\|\phi\|_\infty + m_1\widetilde{P}_\mu\right)\exp\left(S_n\phi(x) - n\widetilde{P}_\mu\right),\end{aligned}$$

and

$$\begin{aligned}\mu(X^n) &\geq \mu\left(B_d\left(p, \Lambda^{-(n+m_1)}\right)\right) \geq \frac{1}{\widetilde{C}_\mu}\exp\left(S_{n+m_1}\phi(x) - (n+m_1)\widetilde{P}_\mu\right) \\ &\geq \frac{1}{\widetilde{C}_\mu \exp\left(m_1\|\phi\|_\infty + m_1\widetilde{P}_\mu\right)}\exp\left(S_n\phi(x) - n\widetilde{P}_\mu\right).\end{aligned}$$

Thus μ is a Gibbs state for f, $\mathcal{C}$, and ϕ, with $P_\mu = \widetilde{P}_\mu$.

To prove the equivalence, we start with the implication from (1) to (2).

We have already shown above that any radial Gibbs state for f, d, and ϕ must be a Gibbs state for f, $\mathcal{C}$, and ϕ.

If we assume that f has no periodic critical point, then there exists a constant $K \in \mathbb{N}$ such that for each $x \in S^2$ and each $n \in \mathbb{N}_0$, the set $U^n(x)$ is a union of at most K distinct n-tiles, i.e.,

$$\operatorname{card}\{Y^n \in \mathbf{X}^n \mid \text{there exists an } n\text{-tile } X^n \in \mathbf{X}^n \text{ with } x \in X^n \text{ and } X^n \cap Y^n \neq \emptyset\} \leq K.$$

Indeed, if f has no periodic critical point, then there exists a constant $N \in \mathbb{N}$ such that $\deg_{f^n}(x) \leq N$ for all $x \in S^2$ and all $n \in \mathbb{N}$ ([BM17, Lemma 18.6]). Since each n-flower $W^n(p)$ for $p \in \mathbf{V}^n$ is covered by exactly $2\deg_{f^n}(p)$ distinct n-tiles ([BM17, Lemma 5.28(i)]), $U^n(x)$ is covered by a bounded number of n-flowers and thus covered by a bounded number, independent of $x \in S^2$ and $n \in \mathbb{N}_0$, of distinct n-tiles.

If μ is a Gibbs state with constants P_μ and C_μ, then by (5.43) and Lemma 5.3, for each $n \in \mathbb{N}_0$ and each $x \in S^2$, we have

$$\begin{aligned}\mu\big(B_d(x, \Lambda^{-n})\big) \geq & \mu\left(U^{n+m_2}(x)\right) \geq C_\mu^{-1} \exp(S_{n+m_2}\phi(x) - (n+m_2)P_\mu) \\ \geq & \frac{1}{C_\mu \exp(m_2 \|\phi\|_\infty + m_2 P_\mu)} \exp(S_n\phi(x) - nP_\mu),\end{aligned}$$

and moreover, if $n \geq m_2$, then

$$\begin{aligned}&\mu\big(B_d(x, \Lambda^{-n})\big) \leq \mu\left(U^{n-m_2}(x)\right) \leq \sum_{\substack{X \in \mathbf{X}^{n-m_2} \\ X \subseteq U^{n-m_2}(x)}} \mu(X) \\ \leq & K C_\mu \exp\left(2C_1 \left(\operatorname{diam}_d(S^2)\right)^\alpha\right) \exp(S_{n-m_2}\phi(x) - (n-m_2)P_\mu) \\ \leq & K C_\mu \exp\left(2C_1 \left(\operatorname{diam}_d(S^2)\right)^\alpha + m_2\left(\|\phi\|_\infty + P_\mu\right)\right) \exp(S_n\phi(x) - nP_\mu),\end{aligned}$$

and if $n < m_2$, then

$$\mu(B_d(x, \Lambda^{-n})) \leq 1 \leq \exp\left(m_2(\|\phi\|_\infty + P_\mu)\right) \exp(S_n\phi(x) - nP_\mu).$$

Thus μ is a radial Gibbs state for f, d, and ϕ.

Next, we show that (2) implies (3).

We assume (2) now. Let $\mu = m_\phi$, where m_ϕ is from Theorem 5.12. Then by Proposition 5.14, μ is a Gibbs state for f, $\mathscr{C}$, and ϕ. Thus μ is also a radial Gibbs state for f, d, and ϕ.

Finally, we prove the implication from (3) to (1) by contradiction.

Assume that f has a periodic critical point $x \in S^2$ with a period $l \in \mathbb{N}$, and let μ be a radial Gibbs state for f, d, and ϕ with constants $\widetilde{P}_\mu$ and $\widetilde{C}_\mu$. So μ is also a Gibbs state for f, $\mathscr{C}$, and ϕ with constants $P_\mu = \widetilde{P}_\mu$ and C_μ, as shown in the first part of the proof.

We note that $x \in \operatorname{post} f \subseteq \mathbf{V}^n$ for each $n \in \mathbb{N}_0$. By (2.5), (2.7), and (5.43), for each $n \in \mathbb{N}$,

$$\overline{W}^{nl+m_2}(x) \subseteq U^{nl+m_2}(x) \subseteq B_d(x, \Lambda^{-nl}).$$

Recall that the number of distinct $(nl+m_2)$-tiles contained in $\overline{W}^{nl+m_2}(x)$ is $2\deg_{f^{nl+m_2}}(x)$. Denote these $(nl+m_2)$-tiles by $X_i^{nl+m_2} \in \mathbf{X}^{nl+m_2}$, $i \in \{1, 2, \dots, 2\deg_{f^{nl+m_2}}(x)\}$. Then by Lemma 5.11, there exists an $(nl+m_2+M)$-tile $Y_i \in$

$\mathbf{X}^{nl+m_2+M}$ such that $Y_i \subseteq \operatorname{inte}\left(X_i^{nl+m_2}\right)$. Here $M \in \mathbb{N}$ is a constant from Lemma 5.11. We fix $x_i \in Y_i$ for each $i \in \{1, 2, \dots, 2\deg_{f^{nl+m_2}}(x)\}$. Note that $Y_i \cap Y_j = \emptyset$ for $1 \le i < j \le 2\deg_{f^{nl+m_2}}(x)$. Thus

$$
\begin{aligned}
&\widetilde{C}_\mu \exp(S_{nl}\phi(x) - nlP_\mu) \ge \mu\big(B_d(x, \Lambda^{-nl})\big) \\
\ge&\mu\left(\overline{W}^{nl+m_2}(x)\right) \ge \sum_{i=1}^{2\deg_{f^{nl+m_2}}(x)} \mu(Y_i) \\
\ge&2\deg_{f^{nl+m_2}}(x)\frac{1}{C_\mu}\exp\left(S_{nl+m_2+M}\phi(x_i) - (nl+m_2+M)P_\mu\right) \\
\ge&\frac{2\left(\deg_{f^l}(x)\right)^n}{C_\mu \exp(M\|\phi\|_\infty + MP_\mu)}\exp(S_{nl+m_2}\phi(x_i) - (nl+m_2)P_\mu) \\
\ge&\frac{2\left(\deg_{f^l}(x)\right)^n \exp(S_{nl+m_2}\phi(x) - (nl+m_2)P_\mu)}{C_\mu \exp\left(M\|\phi\|_\infty + MP_\mu + C_1\left(\operatorname{diam}_d(S^2)\right)^\alpha\right)} \\
\ge&\frac{\left(\deg_{f^l}(x)\right)^n \exp(S_{nl}\phi(x) - nlP_\mu)}{C_\mu \exp\left((m_2+M)\|\phi\|_\infty + (m_2+M)P_\mu + C_1\left(\operatorname{diam}_d(S^2)\right)^\alpha\right)},
\end{aligned}
$$

where the second-to-last inequality follows from Lemma 5.3 and the fact that $x_i, x \in X_i^{nl+m_2}$ for $i \in \{1, 2, \dots, 2\deg_{f^{nl+m_2}}(x)\}$, and C_1 is a constant from Lemma 5.3. So

$$
\left(\deg_{f^l}(x)\right)^n \le \widetilde{C}_\mu C_\mu \exp\left((m_2+M)(\|\phi\|_\infty + P_\mu) + C_1\left(\operatorname{diam}_d(S^2)\right)^\alpha\right),
$$

for each $n \in \mathbb{N}$. However, since x is a periodic critical point of f, we have $\deg_{f^l}(x) > 1$, a contradiction. □

As an immediate consequence, we get that if the expanding Thurston map does not have periodic critical points, then the property of being a Gibbs state does not depend on the choice of the Jordan curve $\mathscr{C} \subseteq S^2$ that satisfies the Assumptions.

Corollary 5.22 *Let f, d, ϕ, α satisfy the Assumptions. We assume that f does not have periodic critical points. Let $\mathscr{C}_1$ and $\mathscr{C}_2$ be Jordan curves on S^2 that satisfy the Assumptions for $\mathscr{C}$, and $\mu \in \mathscr{P}(S^2)$ be a Borel probability measure. Then μ is a Gibbs state with respect to f, C_1, and ϕ if and only if μ is a Gibbs state with respect to f, C_2, and ϕ.*

Proof By Proposition 5.21, since f does not have periodic critical points, f is a Gibbs state with respect to f, $\mathscr{C}_1$, and ϕ if and only if f is a radial Gibbs state with respect to f, d, and ϕ if and only if f is a Gibbs state with respect to f, $\mathscr{C}_2$, and ϕ. □

5.3 Uniqueness

To prove the uniqueness of the equilibrium state of a continuous map g on a compact metric space X, one of the techniques is to prove the (Gâteaux) differentiability of the topological pressure function $P(g,\cdot)\colon C(X) \to \mathbb{R}$. We summarize the general ideas below, but refer the reader to [PU10, Sect. 3.6] for a detailed treatment in the case of forward-expansive maps and distance expanding maps.

For a continuous map $g\colon X \to X$ on a compact metric space X, the topological pressure function $P(g,\cdot)!: C(X) \to \mathbb{R}$ is Lipschitz continuous ([PU10, Theorem 3.6.1]) and convex ([PU10, Theorem 3.6.2]). For an arbitrary convex continuous function $Q\colon V \to \mathbb{R}$ on a real topological vector space V, we call a continuous linear functional $L\colon V \to \mathbb{R}$ *tangent to* Q *at* $x \in V$ if

$$Q(x) + L(y) \leq Q(x+y), \qquad \text{for each } y \in V. \tag{5.44}$$

We denote the set of all continuous linear functionals tangent to Q at $x \in V$ by $V^*_{x,Q}$. It is known (see for example, [PU10, Proposition 3.6.6]) that if $\mu \in \mathcal{M}(X,g)$ is an equilibrium state for g and $\phi \in C(X)$, then the continuous linear functional $u \longmapsto \int u \,\mathrm{d}\mu$ for $u \in C(X)$ is tangent to the topological pressure function $P(g,\cdot)$ at ϕ. Indeed, let $\phi, \gamma \in C(X)$ and $\mu \in \mathcal{M}(X,g)$ be an equilibrium state for g and ϕ. Then $P(g,\phi+\gamma) \geq h_\mu(g) + \int \phi + \gamma \,\mathrm{d}\mu$ by the Variational Principle (3.5), and $P(g,\phi) = h_\mu(g) + \int \phi \,\mathrm{d}\mu$. It follows that $P(g,\phi) + \int \gamma \,\mathrm{d}\mu \leq P(g,\phi+\gamma)$.

Thus in order to prove the uniqueness of the equilibrium state for an expanding Thurston map $f\colon S^2 \to S^2$ and a real-valued Hölder continuous potential ϕ, it suffices to prove that $\operatorname{card}\big(V^*_{\phi,P(f,\cdot)}\big) = 1$. Then we can apply the following fact from functional analysis (see [PU10, Theorem 3.6.5] for a proof):

Theorem 5.23 *Let V be a separable Banach space and $Q\colon V \to \mathbb{R}$ be a convex continuous function. Then for each $x \in V$, the following statements are equivalent:*

1. $\operatorname{card}\big(V^*_{x,Q}\big) = 1$.
2. *The function $t \longmapsto Q(x+ty)$ is differentiable at 0 for each $y \in V$.*
3. *There exists a subset $U \subseteq V$ that is dense in the weak topology on V such that the function $t \longmapsto Q(x+ty)$ is differentiable at 0 for each $y \in U$.*

Now the problem of the uniqueness of equilibrium states transforms to the problem of (Gâteaux) differentiability of the topological pressure function. To investigate the latter, we need a closer study of the fine properties of the Ruelle operator $\mathcal{L}_\phi$.

Let (X,d) be a metric space. A function $h\colon [0,+\infty) \to [0,+\infty)$ is an *abstract modulus of continuity* if it is continuous at 0, non-decreasing, and $h(0)=0$. Given any constant $b \in [0,+\infty]$, and any abstract modulus of continuity h, we define the subclass $C^b_h((X,d),\mathbb{C})$ of $C(X,\mathbb{C})$ as

$$C^b_h((X,d),\mathbb{C}) = \{u \in C(X,\mathbb{C}) \mid \|u\|_\infty \leq b \text{ and for } x,y \in X,\ |u(x)-u(y)| \leq h(d(x,y))\}.$$

We denote $C_h^b(X, d) = C_h^b((X, d), \mathbb{C}) \cap C(X)$.

Assume now that (X, d) is compact. Then by the Arzelà-Ascoli Theorem, each $C_h^b((X, d), \mathbb{C})$ (resp. $C_h^b(X, d)$) is precompact in $C(X, \mathbb{C})$ (resp. $C(X)$) equipped with the uniform norm. It is easy to see that each $C_h^b((X, d), \mathbb{C})$ (resp. $C_h^b(X, d)$) is actually compact. On the other hand, for $u \in C(X, \mathbb{C})$, we can define an abstract modulus of continuity by

$$h(t) = \sup\{|u(x) - u(y)| \mid x, y \in X, d(x, y) \leq t\} \tag{5.45}$$

for $t \in [0, +\infty)$, so that $u \in C_h^\beta((X, d), \mathbb{C})$, where $\beta = \|u\|_\infty$.

We will need the following lemma in this section.

Lemma 5.24 *Let (X, d) be a metric space. For each pair of constants $b_1, b_2 \geq 0$ and each pair of abstract moduli of continuity h_1, h_2, there exist a constant $b \geq 0$ and an abstract modulus of continuity h such that*

$$\left\{u_1 u_2 \,\middle|\, u_1 \in C_{h_1}^{b_1}((X, d), \mathbb{C}), u_2 \in C_{h_2}^{b_2}((X, d), \mathbb{C})\right\} \subseteq C_h^b((X, d), \mathbb{C}), \tag{5.46}$$

and for each $c > 0$,

$$\left\{\frac{1}{u} \,\middle|\, u \in C_{h_1}^{b_1}((X, d), \mathbb{C}), u(x) \geq c \text{ for each } x \in X\right\} \subseteq C_{c^{-2}h_1}^{c^{-1}}((X, d), \mathbb{C}). \tag{5.47}$$

More precisely, we can choose $b = b_1 b_2$ and $h = b_1 h_2 + b_2 h_1$.

Proof For $u_1 \in C_{h_1}^{b_1}((X, d), \mathbb{C})$, $u_2 \in C_{h_2}^{b_2}((X, d), \mathbb{C})$, we have $\|u_1 u_2\|_\infty \leq b_1 b_2$, and for $x, y \in X$,

$$\begin{aligned} |u_1(x)u_2(x) - u_1(y)u_2(y)| &\leq |u_1(x)|\,|u_2(x) - u_2(y)| + |u_2(y)|\,|u_1(x) - u_1(y)| \\ &\leq b_1 h_2(d(x, y)) + b_2 h_1(d(x, y)). \end{aligned}$$

For $c > 0$ and $u \in C_{h_1}^{b_1}((X, d), \mathbb{C})$ with $u(x) \geq c$ for each $x \in X$, we have $\left\|\frac{1}{u}\right\|_\infty \leq \frac{1}{c}$, and for $x, y \in X$,

$$\left|\frac{1}{u(x)} - \frac{1}{u(y)}\right| \leq \left|\frac{u(x) - u(y)}{u(x)u(y)}\right| \leq \frac{1}{c^2} h_1(d(x, y)).$$

□

Let f, d, ϕ, α satisfy the Assumptions. Recall that by (5.27) and Proposition 5.19,

$$\overline{\phi} = \phi - P(f, \phi).$$

We define the function

$$\widetilde{\phi} = \phi - P(f, \phi) + \log u_\phi - \log(u_\phi \circ f), \tag{5.48}$$

where u_ϕ is the continuous function given by Theorem 5.17. Then for $u \in C(S^2)$ and $x \in S^2$, we have

$$\begin{aligned}\mathcal{L}_{\widetilde{\phi}}(u)(x) &= \sum_{y \in f^{-1}(x)} \deg_f(y) u(y) \exp(\phi(y) - P(f, \phi) + \log u_\phi(y) - \log u_\phi(f(y))) \\ &= \frac{1}{u_\phi(x)} \sum_{y \in f^{-1}(x)} \deg_f(y) u(y) u_\phi(y) \exp(\phi(y) - P(f, \phi)) \\ &= \frac{1}{u_\phi(x)} \mathcal{L}_{\overline{\phi}}(u u_\phi)(x),\end{aligned}$$

and thus

$$\mathcal{L}^n_{\widetilde{\phi}}(u)(x) = \frac{1}{u_\phi(x)} \mathcal{L}^n_{\overline{\phi}}(u u_\phi)(x), \qquad \text{for } n \in \mathbb{N}. \tag{5.49}$$

Recall m_ϕ from Theorem 5.12. Then we can show that $\mu_\phi = u_\phi m_\phi$ satisfies

$$\mathcal{L}^*_{\widetilde{\phi}}(\mu_\phi) = \mu_\phi. \tag{5.50}$$

Indeed, by (5.49) and (5.30), for $u \in C(S^2)$,

$$\begin{aligned}\int u \,\mathrm{d}\big(\mathcal{L}^*_{\widetilde{\phi}}(\mu_\phi)\big) &= \int \mathcal{L}_{\widetilde{\phi}}(u) u_\phi \,\mathrm{d}m_\phi = \int \mathcal{L}_{\overline{\phi}}(u u_\phi) \,\mathrm{d}m_\phi \\ &= \int u u_\phi \,\mathrm{d}\big(\mathcal{L}^*_{\overline{\phi}}(m_\phi)\big) = \int u u_\phi \,\mathrm{d}m_\phi = \int u \,\mathrm{d}\mu_\phi.\end{aligned}$$

By (5.34) and (5.49), $\mathcal{L}_{\widetilde{\phi}}(\mathbb{1}) = \frac{1}{u_\phi} \mathcal{L}_{\overline{\phi}}(u_\phi) = \mathbb{1}$, i.e.,

$$\sum_{y \in f^{-1}(x)} \deg_f(y) \exp \widetilde{\phi}(y) = 1 \qquad \text{for } x \in S^2. \tag{5.51}$$

Lemma 5.25 *Let f, d, ϕ, α satisfy the Assumptions. Then the operator norm of $\mathcal{L}_{\widetilde{\phi}}$ acting on $C(S^2)$ is given by $\|\mathcal{L}_{\widetilde{\phi}}\| = 1$. In addition, $\mathcal{L}_{\widetilde{\phi}}(\mathbb{1}) = \mathbb{1}$.*

Proof By (5.51), for each $x \in S^2$ and each $u \in C(S^2)$, we have

$$\begin{aligned}\big|\mathcal{L}_{\widetilde{\phi}}(u)(x)\big| &= \Bigg| \sum_{y \in f^{-1}(x)} \deg_f(y) u(y) \exp \widetilde{\phi}(y) \Bigg| \\ &\leq \|u\|_\infty \sum_{y \in f^{-1}(x)} \deg_f(y) \exp \widetilde{\phi}(y) \\ &= \|u\|_\infty .\end{aligned}$$

Thus $\|\mathcal{L}_{\widetilde{\phi}}\| \leq 1$. Since $\mathcal{L}_{\widetilde{\phi}}(\mathbb{1}) = \mathbb{1}$ by (5.51), $\|\mathcal{L}_{\widetilde{\phi}}\| = 1$. □

Lemma 5.26 *Let f, d, ϕ, α satisfy the Assumptions. Then*

$$\widetilde{\phi} \in C^{0,\alpha}(S^2, d). \tag{5.52}$$

Proof We fix a Jordan curve $\mathscr{C} \subseteq S^2$ that satisfies the Assumptions (see Theorem 2.16 for the existence of such $\mathscr{C}$). By Theorem 5.17, $u_\phi \in C^{0,\alpha}(S^2, d)$ and $C_2^{-1} \leq u_\phi(x) \leq C_2$ for each $x \in S^2$, where $C_2 \geq 1$ is a constant from Lemma 5.4. So $\log u_\phi \in C^{0,\alpha}(S^2, d)$. Note that $\phi \in C^{0,\alpha}(S^2, d)$, so by (5.48) it suffices to prove that $u_\phi \circ f \in C^{0,\alpha}(S^2, d)$. But this follows from the fact that f is Lipschitz with respect to d (Lemma 2.15) and $u_\phi \in C^{0,\alpha}(S^2, d)$. □

Theorem 5.27 *Let $f, \mathscr{C}, d, \Lambda, L$ satisfy the Assumptions. Then for each $\alpha \in (0, 1]$, each $b \geq 0$, and each $\theta \geq 0$, there exist constants $\widehat{b} \geq 0$ and $\widehat{C} \geq 0$ with the following property:*

For each abstract modulus of continuity h, there exists an abstract modulus of continuity $\widetilde{h}$ such that for each $\phi \in C^{0,\alpha}(S^2, d)$ with $|\phi|_\alpha \leq \theta$, we have

$$\left\{\mathscr{L}_{\overline{\phi}}^n(u) \mid u \in C_h^b(S^2, d), n \in \mathbb{N}_0\right\} \subseteq C_{\widehat{h}}^{\widehat{b}}(S^2, d), \tag{5.53}$$

$$\left\{\mathscr{L}_{\widetilde{\phi}}^n(u) \mid u \in C_h^b(S^2, d), n \in \mathbb{N}_0\right\} \subseteq C_{\widetilde{h}}^{b}(S^2, d), \tag{5.54}$$

where $\widehat{h}(t) = \widehat{C}(t^\alpha + h(2C_0 Lt))$ is an abstract modulus of continuity, and $C_0 > 1$ is a constant depending only on $f, \mathscr{C}$, and d from Lemma 2.19.

Proof Fix arbitrary $\alpha \in (0, 1]$, $b \geq 0$, and $\theta \geq 0$. By Lemma 5.16, for $n \in \mathbb{N}_0$, $u \in C_h^b(S^2, d)$, and $\phi \in C^{0,\alpha}(S^2, d)$ with $|\phi|_\alpha \leq \theta$, we have

$$\left\|\mathscr{L}_{\overline{\phi}}^n(u)\right\|_\infty \leq \|u\|_\infty \left\|\mathscr{L}_{\overline{\phi}}^n(\mathbb{1})\right\|_\infty \leq C_2 \|u\|_\infty,$$

where C_2 is the constant defined in (5.7) in Lemma 5.4. So we can choose $\widehat{b} = C_2 b$. Note that by (5.7) that C_2 only depends on $f, \mathscr{C}, d, \theta$, and α.

Let X^0 be either the white 0-tile $X_w^0 \in \mathbf{X}^0(f, \mathscr{C})$ or the black 0-tile $X_b^0 \in \mathbf{X}^0(f, \mathscr{C})$. If $X^m \in \mathbf{X}^m(f, \mathscr{C})$ is an m-tile with $f^m(X^m) = X^0$ for some $m \in \mathbb{N}_0$, then by Proposition 2.6(i), the restriction $f^m|_{X^m}$ of f^m to X^m is a bijection from X^m to X^0. So for each $x \in X^0$, there exists a unique point contained in X^m whose image under f^m is x. We denote this unique point by $x_m(X^m)$. Note that for each $z = x_m(X^m)$, the number of distinct m-tiles $X \in \mathbf{X}^m(f, \mathscr{C})$ that satisfy both $f^m(X) = X^0$ and $x_m(X) = z$ is exactly $\deg_{f^m}(z)$.

Then for each $x, y \in X^0$, we have

$$
\begin{aligned}
&\left|\mathscr{L}^n_{\overline{\phi}}(u)(x)-\mathscr{L}^n_{\overline{\phi}}(u)(y)\right| \\
=&\left|\sum_{\substack{X^n\in\mathbf{X}^n(f,\mathscr{C})\\ f^n(X^n)=X^0}}(u\exp(S_n\overline{\phi}))(x_n(X^n))-(u\exp(S_n\overline{\phi}))(y_n(X^n))\right| \\
\leq&\left|\sum_{\substack{X^n\in\mathbf{X}^n(f,\mathscr{C})\\ f^n(X^n)=X^0}}u(x_n(X^n))\left(\exp(S_n\overline{\phi}(x_n(X^n)))-\exp(S_n\overline{\phi}(y_n(X^n)))\right)\right| \\
&+\left|\sum_{\substack{X^n\in\mathbf{X}^n(f,\mathscr{C})\\ f^n(X^n)=X^0}}\exp(S_n\overline{\phi}(y_n(X^n)))\left(u(x_n(X^n))-u(y_n(X^n))\right)\right|.
\end{aligned}
$$

The second term above is

$$
\leq C_2h(C_0\Lambda^{-n}d(x,y))\leq C_2h(C_0d(x,y)),
$$

due to (5.32) and the fact that $d(x_n(X^n),y_n(X^n))\leq C_0\Lambda^{-n}d(x,y)$ by Lemma 2.19, where the constant C_0 comes from.

In order to bound the first term, we define

$$
A_n^+=\{X^n\in\mathbf{X}^n(f,\mathscr{C})\mid f^n(X^n)=X^0,\ S_n\overline{\phi}(x_n(X^n))\geq S_n\overline{\phi}(y_n(X^n))\},
$$

and

$$
A_n^-=\{X^n\in\mathbf{X}^n(f,\mathscr{C})\mid f^n(X^n)=X^0,\ S_n\overline{\phi}(x_n(X^n))<S_n\overline{\phi}(y_n(X^n))\}.
$$

Then by (3.8), Lemma 5.3, and Lemma 5.16 the first term is

$$
\begin{aligned}
&\leq\sum_{X^n\in A_n^+}\|u\|_\infty\left(\exp(S_n\overline{\phi}(x_n(X^n)))-\exp(S_n\overline{\phi}(y_n(X^n)))\right) \\
&\quad+\sum_{X^n\in A_n^-}\|u\|_\infty\left(\exp(S_n\overline{\phi}(y_n(X^n)))-\exp(S_n\overline{\phi}(x_n(X^n)))\right) \\
&=\|u\|_\infty\left(\left(\frac{\sum\limits_{X^n\in A_n^+}\exp(S_n\overline{\phi}(x_n(X^n)))}{\sum\limits_{X^n\in A_n^+}\exp(S_n\overline{\phi}(y_n(X^n)))}-1\right)\sum_{X^n\in A_n^+}e^{S_n\overline{\phi}(y_n(X^n))}\right. \\
&\quad\left.+\left(\frac{\sum\limits_{X^n\in A_n^-}\exp(S_n\overline{\phi}(y_n(X^n)))}{\sum\limits_{X^n\in A_n^-}\exp(S_n\overline{\phi}(x_n(X^n)))}-1\right)\sum_{X^n\in A_n^-}e^{S_n\overline{\phi}(x_n(X^n))}\right) \\
&\leq 2bC_2(\exp(C_1d(x,y)^\alpha)-1) \\
&\leq 2b\widetilde{C}_3d(x,y)^\alpha,
\end{aligned}
$$

for some constant $\widetilde{C}_3 > 0$ that only depends on C_1, C_2, and $\operatorname{diam}_d(S^2)$, where $C_1 > 0$ is the constant defined in (5.5) in Lemma 5.3 and $C_2 \geq 1$ is the constant defined in (5.7) in Lemma 5.4. Note that the justification of the second inequality above is similar to that of (5.33) in Lemma 5.16. We observe that by (5.5) and (5.7), both C_1 and C_2 only depend on $f, \mathscr{C}, d, \theta$, and α, so does $\widetilde{C}_3$.

Hence we get

$$\left|\mathscr{L}_{\overline{\phi}}^n(u)(x) - \mathscr{L}_{\overline{\phi}}^n(u)(y)\right| \leq 2b\widetilde{C}_3 d(x,y)^\alpha + C_2 h(C_0 d(x,y)).$$

Now we consider arbitrary $x \in X_w^0$ and $y \in X_b^0$. Since the metric space (S^2, d) is linearly locally connected with a linear local connectivity constant $L \geq 1$, there exists a continuum $E \subseteq S^2$ with $x, y \in E$ and $E \subseteq B_d(x, Ld(x,y))$. We can then choose $z \in \mathscr{C} \cap E$. Note that $\max\{d(x,z), d(y,z)\} \leq 2Ld(x,y)$.

Hence we get

$$\begin{aligned}
&\left|\mathscr{L}_{\overline{\phi}}^n(u)(x) - \mathscr{L}_{\overline{\phi}}^n(u)(y)\right| \\
\leq &\left|\mathscr{L}_{\overline{\phi}}^n(u)(x) - \mathscr{L}_{\overline{\phi}}^n(u)(z)\right| + \left|\mathscr{L}_{\overline{\phi}}^n(u)(z) - \mathscr{L}_{\overline{\phi}}^n(u)(y)\right| \\
\leq &2b\widetilde{C}_3 d(x,z)^\alpha + C_2 h(C_0 d(x,z)) + 2b\widetilde{C}_3 d(z,y)^\alpha + C_2 h(C_0 d(z,y)) \\
\leq &8bL\widetilde{C}_3 d(x,y)^\alpha + 2C_2 h(2C_0 L d(x,y)).
\end{aligned}$$

By choosing $\widehat{C} = \max\left\{8bL\widetilde{C}_3, 2C_2\right\}$, which only depends on $f, \mathscr{C}, d, \theta$, and α, we complete the proof of (5.53).

We now prove (5.54).

We fix an arbitrary $\phi \in C^{0,\alpha}(S^2, d)$ with $|\phi|_\alpha \leq \theta$. Then by (5.35) in Theorem 5.17 and (5.7) in Lemma 5.4, we have

$$\left\|u_\phi\right\|_\infty \leq b_1,$$

where $b_1 = \exp\left(4\frac{\theta C_0}{1-\Lambda^{-1}} L\left(\operatorname{diam}(S^2)\right)^\alpha\right)$. By Theorem 5.17 and (5.33) in Lemma 5.16, for each $x, y \in S^2$, we have

$$\begin{aligned}
\left|u_\phi(x) - u_\phi(y)\right| &= \left|\lim_{n\to+\infty} \frac{1}{n}\sum_{j=0}^{n-1}\left(\mathscr{L}_{\overline{\phi}}^j(\mathbb{1})(x) - \mathscr{L}_{\overline{\phi}}^j(\mathbb{1})(x)\right)\right| \\
&\leq \limsup_{n\to+\infty} \frac{1}{n}\sum_{j=0}^{n-1}\left|\mathscr{L}_{\overline{\phi}}^j(\mathbb{1})(x) - \mathscr{L}_{\overline{\phi}}^j(\mathbb{1})(x)\right| \\
&\leq C_2\left(\exp\left(4C_1 L d(x,y)^\alpha\right) - 1\right).
\end{aligned}$$

So

$$u_\phi \in C_{h_1}^{b_1}(S^2, d), \tag{5.55}$$

where h_1 is an abstract modulus of continuity given by

$$h_1(t) = C_2 \left(\exp\left(4C_1 L t^{\alpha}\right) - 1\right), \quad \text{for } t \in [0, +\infty).$$

Thus by Lemma 5.24, there exist a constant $b_2 \geq 0$ and an abstract modulus of continuity h_2 such that

$$\left\{uu_{\phi} \,\middle|\, u \in C_h^b(S^2, d), \phi \in C^{0,\alpha}(S^2, d), |\phi|_{\alpha} \leq \theta\right\} \subseteq C_{h_2}^{b_2}(S^2, d). \tag{5.56}$$

Then by (5.49), (5.56), (5.53), and Lemma 5.24, we get that there exist a constant $b_3 \geq 0$ and an abstract modulus of continuity $\widetilde{h}$ such that

$$\left\{\mathcal{L}_{\widetilde{\phi}}^n(u) \,\middle|\, u \in C_h^b(S^2, d), n \in \mathbb{N}_0\right\} \subseteq C_{\widetilde{h}}^{b_3}(S^2, d), \tag{5.57}$$

for each $\phi \in C^{0,\alpha}(S^2, d)$ with $|\phi|_{\alpha} \leq \theta$.

On the other hand, by Lemma 5.25, $\left\|\mathcal{L}_{\widetilde{\phi}}^n(u)\right\|_{\infty} \leq \|u\|_{\infty} \leq b$ for each $u \in C_h^b(S^2, d)$, each $n \in \mathbb{N}_0$, and each $\phi \in C^{0,\alpha}(S^2, d)$. Therefore, we have proved (5.54). □

As a consequence, both $\mathcal{L}_{\widetilde{\phi}}$ and $\mathcal{L}_{\widetilde{\phi}}$ are almost periodic.

Definition 5.28 A bounded linear operator $L \colon B \to B$ on a Banach space B is *almost periodic* if for each $z \in B$, the closure of the set $\{L^n(z) \mid n \in \mathbb{N}_0\}$ is compact in the norm topology.

Corollary 5.29 *Let f, d, ϕ, and α satisfy the Assumptions. Let $C(S^2)$ be equipped with the uniform norm. Then both $\mathcal{L}_{\widetilde{\phi}} \colon C(S^2) \to C(S^2)$ and $\mathcal{L}_{\widetilde{\phi}} \colon C(S^2) \to C(S^2)$ are almost periodic.*

Proof Fix a Jordan curve $\mathscr{C} \subseteq S^2$ that satisfies the Assumptions (see Theorem 2.16 for the existence of such $\mathscr{C}$). For each $u \in C(S^2)$, we have $u \in C_h^{\beta}(S^2, d)$ for $\beta = \|u\|_{\infty}$ and some abstract modulus of continuity h defined in (5.45). Then the corollary follows immediately from Theorem 5.27 and Arzelà-Ascoli theorem. □

Lemma 5.30 *Let f and d satisfy the Assumptions. Let g be an abstract modulus of continuity. Then for $\alpha \in (0, 1]$, $K \in (0, +\infty)$, and $\delta_1 \in (0, +\infty)$, there exist constants $\delta_2 \in (0, +\infty)$ and $n \in \mathbb{N}$ with the following property:*

For each $u \in C_g^{+\infty}(S^2, d)$, each $\phi \in C^{0,\alpha}(S^2, d)$, and each choice of m_{ϕ} from Theorem 5.12, if $\|\phi\|_{C^{0,\alpha}} \leq K$, $\int uu_{\phi} \, \mathrm{d}m_{\phi} = 0$, and $\|u\|_{\infty} \geq \delta_1$, then

$$\left\|\mathcal{L}_{\widetilde{\phi}}^n(u)\right\|_{\infty} \leq \|u\|_{\infty} - \delta_2.$$

Note that at this point, we have not proved yet that m_{ϕ} from Theorem 5.12 is unique. We will prove it in Corollary 5.32. Recall that u_{ϕ} is the continuous function defined in Theorem 5.17 that only depends on f and ϕ.

Proof Fix arbitrary constants $\alpha \in (0, 1]$, $K \in (0, +\infty)$, and $\delta_1 \in (0, +\infty)$. Fix $\varepsilon > 0$ small enough such that $g(\varepsilon) < \frac{\delta_1}{2}$. Fix a choice of m_ϕ, an arbitrary $\phi \in C^{0,\alpha}(S^2, d)$, and an arbitrary $u \in C_g^{+\infty}(S^2, d)$ with $\|\phi\|_{C^{0,\alpha}} \leq K$, $\int u u_\phi \,\mathrm{d}m_\phi = 0$, and $\|u\|_\infty \geq \delta_1$.

We pick a Jordan curve $\mathscr{C} \subseteq S^2$ that satisfies the Assumptions (see Theorem 2.16 for the existence of such $\mathscr{C}$).

By Lemma 2.13(iv), there exists $n \in \mathbb{N}$ depending only on $f, \mathscr{C}, d, g$, and δ_1 such that for each $z \in S^2$, we have $U^n(z) \subseteq B_d(z, \varepsilon)$, where $U^n(z)$ is defined in (2.7). We may assume that $\mathscr{C}$ was chosen in such a way that n is minimal for given f, d, g, and δ_1. Thus we can choose n as above that only depends on f, d, g, and δ_1.

Since $\int u u_\phi \,\mathrm{d}m_\phi = 0$, there exist points $y_1, y_2 \in S^2$ such that $u(y_1) \leq 0$ and $u(y_2) \geq 0$.

We fix a point $x \in S^2$. Since $f^n(U^n(y_1)) = S^2$, there exists $y \in f^{-n}(x)$ such that $y \in U^n(y_1) \subseteq B_d(y_1, \varepsilon)$. Thus

$$u(y) \leq u(y_1) + g(\varepsilon) < \frac{\delta_1}{2} \leq \|u\|_\infty - \frac{\delta_1}{2}.$$

So by Lemma 5.25 and (3.8) we have

$$\begin{aligned}
&\mathscr{L}_{\widetilde{\phi}}^n(u)(x) \\
&= \deg_{f^n}(y) u(y) \exp\big(S_n\widetilde{\phi}(y)\big) + \sum_{w \in f^{-n}(x) \setminus \{y\}} \deg_{f^n}(w) u(w) \exp\big(S_n\widetilde{\phi}(w)\big) \\
&\leq \left(\|u\|_\infty - \frac{\delta_1}{2}\right) \deg_{f^n}(y) \exp\big(S_n\widetilde{\phi}(y)\big) + \|u\|_\infty \sum_{w \in f^{-n}(x) \setminus \{y\}} \deg_{f^n}(w) \exp\big(S_n\widetilde{\phi}(w)\big) \\
&\leq \|u\|_\infty \sum_{w \in f^{-n}(x)} \deg_{f^n}(w) \exp\big(S_n\widetilde{\phi}(w)\big) - \frac{\delta_1}{2} \exp\big(S_n\widetilde{\phi}(y)\big) \\
&= \|u\|_\infty - \frac{\delta_1}{2} \exp\big(S_n\widetilde{\phi}(y)\big).
\end{aligned}$$

Similarly, there exists $z \in f^{-n}(x)$ such that $z \in U^n(y_2) \subseteq B_d(y_2, \varepsilon)$ and

$$\mathscr{L}_{\widetilde{\phi}}^n(u)(x) \geq -\|u\|_\infty + \frac{\delta_1}{2} \exp\big(S_n\widetilde{\phi}(z)\big).$$

Hence we get

$$\left\|\mathscr{L}_{\widetilde{\phi}}^n(u)\right\|_\infty \leq \|u\|_\infty - \frac{\delta_1}{2} \inf\big\{\exp\big(S_n\widetilde{\phi}(w)\big) \,\big|\, w \in S^2\big\}. \tag{5.58}$$

Now it suffices to bound each term in the definition of $\widetilde{\phi}$ in (5.48).

First, by the hypothesis, $\|\phi\|_\infty \leq \|\phi\|_{C^{0,\alpha}} \leq K$ (see (0.2)).

Next, for each fixed $x \in S^2$, by Proposition 5.19, we have

$$\begin{aligned} P(f,\phi) &= \lim_{n\to+\infty} \frac{1}{n} \log \sum_{y\in f^{-n}(x)} \deg_{f^n}(y) \exp(S_n\phi(y)) \\ &\leq \lim_{n\to+\infty} \frac{1}{n} \log \sum_{y\in f^{-n}(x)} \deg_{f^n}(y) \exp(nK) \\ &= K + \lim_{n\to+\infty} \frac{1}{n} \log \sum_{y\in f^{-n}(x)} \deg_{f^n}(y) \\ &= K + \log(\deg f). \end{aligned}$$

Similarly, $P(f,\phi) \geq -K + \log(\deg f)$. So $|P(f,\phi)| \leq K + |\log(\deg f)|$.

Finally, by Theorem 5.17 and (5.7) in Lemma 5.4, we have

$$\left\| u_\phi \right\|_\infty \leq C_2 \leq \exp\left(C_5\right),$$

where

$$C_5 = 4\frac{KC_0}{1-\Lambda^{-1}} L \left(\mathrm{diam}_d(S^2)\right)^\alpha,$$

and $C_0 > 1$ is a constant from Lemma 2.19 depending only on f, $\mathscr{C}$, and d.

Therefore, by (5.48) and (5.58), $\left\| \mathscr{L}^n_{\bar{\phi}}(u) \right\|_\infty \leq \|u\|_\infty - \delta_2$, where

$$\delta_2 = \frac{\delta_1}{2} \exp\left(-n\left(2K + |\log(\deg f)| + 2C_5\right)\right),$$

which only depends on f, d, α, K, δ_1, g, and n. □

Theorem 5.31 *Let $f\colon S^2 \to S^2$ be an expanding Thurston map. Let d be a visual metric on S^2 for f with expansion factor $\Lambda > 1$. Let $b \in (0,+\infty)$ be a constant and $h\colon [0,+\infty) \to [0,+\infty)$ an abstract modulus of continuity. Let H be a bounded subset of $C^{0,\alpha}(S^2,d)$ for some $\alpha \in (0,1]$. Then for each $u \in C^b_h(S^2,d)$, each $\phi \in H$, and each choice of m_ϕ from Theorem 5.12, we have*

$$\lim_{n\to+\infty} \left\| \mathscr{L}^n_{\bar{\phi}}(u) - u_\phi \int u \,\mathrm{d}m_\phi \right\|_\infty = 0. \tag{5.59}$$

If, in addition, $\int u u_\phi \,\mathrm{d}m_\phi = 0$, then

$$\lim_{n\to+\infty} \left\| \mathscr{L}^n_{\bar{\phi}}(u) \right\|_\infty = 0. \tag{5.60}$$

Moreover, the convergence in both (5.59) and (5.60) is uniform in $u \in C^b_h(S^2,d)$, $\phi \in H$, and the choice of m_ϕ.

The equation (5.60) demonstrates the contracting behavior of $\mathscr{L}_{\bar{\phi}}$ on a codimension-1 subspace of $C(S^2)$.

Proof Let L be a linear local connectivity constant of d. We can assume that $H \neq \emptyset$. Define a constant $K = \sup\{\|\phi\|_{C^{0,\alpha}} \mid \phi \in H\} \in [0, +\infty)$.

We fix a Jordan curve $\mathscr{C} \subseteq S^2$ that satisfies the Assumptions (see Theorem 2.16 for the existence of such $\mathscr{C}$).

Let M_ϕ be the set of possible choices of m_ϕ from Theorem 5.12, i.e.,

$$M_\phi = \{m \in \mathscr{P}(S^2) \mid \mathscr{L}_\phi^*(m) = cm \text{ for some } c \in \mathbb{R}\}. \tag{5.61}$$

We recall that μ_ϕ defined in Theorem 5.17 by $\mu_\phi = u_\phi m_\phi$ depends on the choice of m_ϕ.

Define for each $n \in \mathbb{N}_0$,

$$a_n = \sup\left\{ \left\|\mathscr{L}_{\overline{\phi}}^n(u)\right\|_\infty \,\middle|\, \phi \in H, u \in C_h^b(S^2, d), \int u \,\mathrm{d}\mu_\phi = 0, m_\phi \in M_\phi \right\}.$$

By Lemma 5.25, $\|\mathscr{L}_{\overline{\phi}}\| = 1$, so $\left\|\mathscr{L}_{\overline{\phi}}^n(u)\right\|_\infty$ is non-increasing in n for fixed $\phi \in H$ and $u \in C_h^b(S^2, d)$. Note that $a_0 \leq b < +\infty$. Thus $\{a_n\}_{n \in \mathbb{N}_0}$ is a non-increasing sequence of non-negative real numbers.

Suppose now that $\lim\limits_{n \to +\infty} a_n = a > 0$. By Theorem 5.27, there exists an abstract modulus of continuity g such that

$$\left\{\mathscr{L}_{\overline{\phi}}^n(u) \,\middle|\, n \in \mathbb{N}_0, \phi \in H, u \in C_h^b(S^2, d)\right\} \subseteq C_g^b(S^2, d).$$

Note that for each $\phi \in H$, each $n \in \mathbb{N}_0$, and each $u \in C_h^b(S^2, d)$ with $\int u u_\phi \,\mathrm{d}m_\phi = 0$, we have $\int \mathscr{L}_{\overline{\phi}}^n(u) u_\phi \,\mathrm{d}m_\phi = \int \mathscr{L}_{\overline{\phi}}^n(u) \,\mathrm{d}\mu_\phi = 0$ by (5.50). So by applying Lemma 5.30 with g, α, K, and $\delta_1 = \frac{a}{2}$, we find constants $n_0 \in \mathbb{N}$ and $\delta_2 > 0$ such that

$$\left\|\mathscr{L}_{\overline{\phi}}^{n_0}\left(\mathscr{L}_{\overline{\phi}}^n(u)\right)\right\|_\infty \leq \left\|\mathscr{L}_{\overline{\phi}}^n(u)\right\|_\infty - \delta_2, \tag{5.62}$$

for each $n \in \mathbb{N}_0$, each $\phi \in H$, each $m_\phi \in M_\phi$, and each $u \in C_h^b(S^2, d)$ with $\int u u_\phi \,\mathrm{d}m_\phi = 0$ and $\left\|\mathscr{L}_{\overline{\phi}}^n(u)\right\|_\infty \geq \frac{a}{2}$. Since $\lim\limits_{n \to +\infty} a_n = a$, we can fix $m > 1$ large enough such that $a_m \leq a + \frac{\delta_2}{2}$. Then for each $\phi \in H$, each $m_\phi \in M_\phi$, and each $u \in C_h^b(S^2, d)$ with $\int u \,\mathrm{d}\mu_\phi = 0$ and $\left\|\mathscr{L}_{\overline{\phi}}^m(u)\right\|_\infty \geq \frac{a}{2}$, we have

$$\left\|\mathscr{L}_{\overline{\phi}}^{n_0+m}(u)\right\|_\infty \leq \left\|\mathscr{L}_{\overline{\phi}}^m(u)\right\|_\infty - \delta_2 \leq a_m - \delta_2 \leq a - \frac{\delta_2}{2}. \tag{5.63}$$

On the other hand, since $\left\|\mathscr{L}_{\overline{\phi}}^n(u)\right\|_\infty$ is non-increasing in n, we have that for each $\phi \in H$, each $m_\phi \in M_\phi$, and each $u \in C_h^b(S^2, d)$ with $\int u \,\mathrm{d}\mu_\phi = 0$ and $\left\|\mathscr{L}_{\overline{\phi}}^m(u)\right\|_\infty < \frac{a}{2}$, the following holds:

$$\left\| \mathcal{L}_{\overline{\phi}}^{n_0+m}(u) \right\|_\infty \leq \left\| \mathcal{L}_{\overline{\phi}}^{m}(u) \right\|_\infty < \frac{a}{2}. \tag{5.64}$$

Thus $a_{n_0+m} \leq \max\left\{a - \frac{\delta_2}{2}, \frac{a}{2}\right\} < a$, contradicting the fact that $\{a_n\}_{n\in\mathbb{N}_0}$ is a non-increasing sequence and the assumption that $\lim\limits_{n\to+\infty} a_n = a$. This proves the uniform convergence in (5.60).

Next, we prove the uniform convergence in (5.59). By Lemma 5.3, Lemma 5.4, Lemma 5.25, and (5.49), for each $u \in C_h^b(S^2, d)$, each $\phi \in H$, and each $m_\phi \in M_\phi$, we have

$$\begin{aligned} &\left\| \mathcal{L}_{\overline{\phi}}^{n}(u) - u_\phi \int u \,\mathrm{d}m_\phi \right\|_\infty \qquad (5.65)\\ &\leq \left\| u_\phi \right\|_\infty \left\| \frac{1}{u_\phi} \mathcal{L}_{\overline{\phi}}^{n}(u) - \int u \,\mathrm{d}m_\phi \right\|_\infty \\ &= \left\| u_\phi \right\|_\infty \left\| \mathcal{L}_{\widetilde{\phi}}^{n}\left(\frac{u}{u_\phi}\right) - \int \frac{u}{u_\phi} \,\mathrm{d}\mu_\phi \right\|_\infty \\ &= \left\| u_\phi \right\|_\infty \left\| \mathcal{L}_{\widetilde{\phi}}^{n}\left(\frac{u}{u_\phi} - \mathbb{1} \int \frac{u}{u_\phi} \,\mathrm{d}\mu_\phi\right) \right\|_\infty . \end{aligned}$$

By (5.35) and (5.7), we have

$$\exp(-C_5) \leq \left\| u_\phi \right\|_\infty \leq \exp(C_5), \tag{5.66}$$

where

$$C_5 = 4 \frac{K C_0}{1 - \Lambda^{-1}} L \left(\operatorname{diam}_d(S^2)\right)^\alpha,$$

and C_0 is a constant from Lemma 2.19 depending only on f, $\mathcal{C}$, and d. Let $v = \frac{u}{u_\phi} - \mathbb{1} \int \frac{u}{u_\phi} \,\mathrm{d}\mu_\phi$. Then v satisfies

$$\|v\|_\infty \leq 2 \left\| \frac{u}{u_\phi} \right\|_\infty \leq 2b \exp(C_5). \tag{5.67}$$

Due to the first inequality in (5.66) and the fact that $u_\phi \in C^{0,\alpha}(S^2, d)$ by Theorem 5.17, we can apply Lemma 5.24 and conclude that there exists an abstract modulus of continuity g of $\frac{u}{u_\phi}$ such that g is independent of the choices of $u \in C_h^b(S^2, d)$, $\phi \in H$, and $m_\phi \in M_\phi$. Thus $v \in C_g^{\widehat{b}}(S^2, d)$, where $\widehat{b} = 2b \exp(C_5)$. Note that $\int v u_\phi \,\mathrm{d}m_\phi = \int v \,\mathrm{d}\mu_\phi = 0$. Finally, we can apply the uniform convergence in (5.60) with $u = v$ to conclude the uniform convergence in (5.59) by (5.65) and (5.66). □

Theorem 5.31 implies in particular the uniqueness of m_ϕ and μ_ϕ.

Corollary 5.32 *Let f, d, ϕ, α satisfy the Assumptions. Then the measure $m_\phi \in \mathcal{P}(S^2)$ defined in Theorem 5.12 is unique, i.e., m_ϕ is the unique Borel probability measure on S^2 that satisfies $\mathcal{L}_\phi^*(m_\phi) = cm_\phi$ for some constant $c \in \mathbb{R}$. Moreover, $\mu_\phi = u_\phi m_\phi$ is the unique Borel probability measure on S^2 that satisfies $\mathcal{L}_{\widetilde{\phi}}^*(\mu_\phi) = \mu_\phi$. In particular, we have $m_{\widetilde{\phi}} = \mu_\phi$.*

Proof Let $m_\phi, \widehat{m}_\phi \in \mathcal{P}(S^2)$ be two measures, both of which arise from Theorem 5.12. Recall that for each $u \in C(S^2)$, there exists some abstract modulus of continuity h such that $u \in C_h^\beta(S^2, d)$, where $\beta = \|u\|_\infty$. Then by (5.59) and (5.35), we see that $\int u \,\mathrm{d}m_\phi = \int u \,\mathrm{d}\widehat{m}_\phi$ for each $u \in C(S^2)$. Thus $m_\phi = \widehat{m}_\phi$.

By (5.50), $\mathcal{L}_{\widetilde{\phi}}^*(\mu_\phi) = \mu_\phi$. Since $\widetilde{\phi} \in C^{0,\alpha}(S^2, d)$ by Lemma 5.26, we get that $\mu_\phi = m_{\widetilde{\phi}}$ and μ_ϕ is the only Borel probability measure on S^2 that satisfies $\mathcal{L}_{\widetilde{\phi}}^*(\mu_\phi) = \mu_\phi$. □

Lemma 5.33 *Let f and d satisfy the Assumptions. Let $b \geq 0$ be a constant and h an abstract modulus of continuity. Let H be a bounded subset of $C^{0,\alpha}(S^2, d)$ for some $\alpha \in (0, 1]$. Then for each $x \in S^2$, each $u \in C_h^b(S^2, d)$, and each $\phi \in H$, we have*

$$\lim_{n\to+\infty} \frac{\frac{1}{n}\sum\limits_{y\in f^{-n}(x)} \deg_{f^n}(y)\,(S_n u(y)) \exp(S_n\phi(y))}{\sum\limits_{y\in f^{-n}(x)} \deg_{f^n}(y) \exp(S_n\phi(y))} = \int u \,\mathrm{d}\mu_\phi. \tag{5.68}$$

Moreover, the convergence is uniform in $x \in S^2$, $u \in C_h^b(S^2, d)$, and $\phi \in H$.

Proof We fix a Jordan curve $\mathcal{C} \subseteq S^2$ that satisfies the Assumptions (see Theorem 2.16 for the existence of such $\mathcal{C}$). By (3.8) and (2.2), for $x \in S^2$, $u \in C_h^b(S^2, d)$, $\phi \in H$, and $n \in \mathbb{N}$,

$$\begin{aligned}
&\frac{\frac{1}{n}\sum\limits_{y\in f^{-n}(x)} \deg_{f^n}(y)\,(S_n u(y)) \exp(S_n\phi(y))}{\sum\limits_{y\in f^{-n}(x)} \deg_{f^n}(y) \exp(S_n\phi(y))} \\
&= \frac{\frac{1}{n}\sum\limits_{j=0}^{n-1}\sum\limits_{y\in f^{-n}(x)} \deg_{f^n}(y) u(f^j(y)) \exp(S_n\phi(y))}{\mathcal{L}_\phi^n(\mathbb{1})(x)} \\
&= \frac{\frac{1}{n}\sum\limits_{j=0}^{n-1}\sum\limits_{z\in f^{-(n-j)}(x)}\sum\limits_{y\in f^{-j}(z)} \deg_{f^{n-j}}(z) \deg_{f^j}(y) u(z) e^{S_j\phi(y)+S_{n-j}\phi(z)}}{\mathcal{L}_\phi^n(\mathbb{1})(x)} \\
&= \frac{\frac{1}{n}\sum\limits_{j=0}^{n-1}\sum\limits_{z\in f^{-(n-j)}(x)} \deg_{f^{n-j}}(z) u(z) \mathcal{L}_\phi^j(\mathbb{1})(z) \exp(S_{n-j}\phi(z))}{\mathcal{L}_\phi^n(\mathbb{1})(x)}
\end{aligned}$$

$$=\frac{\frac{1}{n}\sum_{j=0}^{n-1}\mathcal{L}_{\phi}^{n-j}\Big(u\mathcal{L}_{\phi}^{j}(\mathbb{1})\Big)(x)}{\mathcal{L}_{\phi}^{n}(\mathbb{1})(x)}$$

$$=\frac{\frac{1}{n}\sum_{j=0}^{n-1}\mathcal{L}_{\overline{\phi}}^{n-j}\Big(u\mathcal{L}_{\overline{\phi}}^{j}(\mathbb{1})\Big)(x)}{\mathcal{L}_{\overline{\phi}}^{n}(\mathbb{1})(x)}.$$

By Theorem 5.27, $\{\mathcal{L}_{\overline{\phi}}^{n}(\mathbb{1}) \,|\, n \in \mathbb{N}_0\} \subseteq C_{\widehat{h}}^{\widehat{b}}(S^2, d)$, for some constant $\widehat{b} \geq 0$ and some abstract modulus of continuity $\widehat{h}$, which are independent of the choice of $\phi \in H$. Thus by Lemma 5.24,

$$\{u\mathcal{L}_{\overline{\phi}}^{n}(\mathbb{1}) \,|\, n \in \mathbb{N}_0, u \in C_h^b(S^2, d)\} \subseteq C_{h_1}^{b_1}(S^2, d), \tag{5.69}$$

for some constant $b_1 \geq 0$ and some abstract modulus of continuity h_1, which are independent of the choice of $\phi \in H$.

Hence, by Theorem 5.31 and Corollary 5.32, we have

$$\left\|\mathcal{L}_{\overline{\phi}}^{l}(\mathbb{1}) - u_\phi\right\|_\infty \longrightarrow 0, \tag{5.70}$$

and

$$\left\|\mathcal{L}_{\overline{\phi}}^{l}\Big(u\mathcal{L}_{\overline{\phi}}^{j}(\mathbb{1})\Big) - u_\phi \int u\mathcal{L}_{\overline{\phi}}^{j}(\mathbb{1})\,\mathrm{d}m_\phi\right\|_\infty \longrightarrow 0, \tag{5.71}$$

as $l \longrightarrow +\infty$, uniformly in $j \in \mathbb{N}_0$, $\phi \in H$, and $u \in C_h^b(S^2, d)$.

Fix a constant $K \in (0, +\infty)$ such that for each $\phi \in H$, $\|\phi\|_{C^{0,\alpha}} \leq K$. By (5.35) and (5.7), we have that for each $x \in S^2$,

$$\exp(-C_5) \leq u_\phi(x) \leq \exp(C_5), \tag{5.72}$$

where

$$C_5 = 4\frac{KC_0}{1-\Lambda^{-1}}L\left(\mathrm{diam}_d(S^2)\right)^\alpha,$$

and $C_0 \geq 1$ is a constant from Lemma 2.19 depending only on f, $\mathscr{C}$, and d. So by (5.69), we get that for $j \in \mathbb{N}_0$, $u \in C_h^b(S^2, d)$, and $\phi \in H$,

$$\left\|u_\phi \int u\mathcal{L}_{\overline{\phi}}^{j}(\mathbb{1})\,\mathrm{d}m_\phi\right\|_\infty \leq \|u_\phi\|_\infty \left\|u\mathcal{L}_{\overline{\phi}}^{j}(\mathbb{1})\right\|_\infty \leq b_1 \exp(C_5). \tag{5.73}$$

By (5.53) in Theorem 5.27 and (5.69), we get some constant $b_2 > 0$ such that for all $j, l \in \mathbb{N}_0$, each $u \in C_h^b(S^2, d)$, and each $\phi \in H$,

$$\left\| \mathcal{L}_{\overline{\phi}}^{l} \left(u \mathcal{L}_{\overline{\phi}}^{j} (\mathbb{1}) \right) \right\|_{\infty} < b_2. \tag{5.74}$$

Hence we can conclude from (5.73), (5.74), and (5.71) that

$$\lim_{n \to +\infty} \frac{1}{n} \left| \sum_{j=0}^{n-1} \mathcal{L}_{\overline{\phi}}^{n-j} \left(u \mathcal{L}_{\overline{\phi}}^{j} (\mathbb{1}) \right) (x) - \sum_{j=0}^{n-1} u_{\phi}(x) \int u \mathcal{L}_{\overline{\phi}}^{j} (\mathbb{1}) \, \mathrm{d} m_{\phi} \right| = 0,$$

uniformly in $u \in C_h^b(S^2, d)$, $\phi \in H$, and $x \in S^2$. Thus by (5.70) and (5.72), we have

$$\lim_{n \to +\infty} \left| \frac{\frac{1}{n} \sum_{j=0}^{n-1} \mathcal{L}_{\overline{\phi}}^{n-j} \left(u \mathcal{L}_{\overline{\phi}}^{j} (\mathbb{1}) \right) (x)}{\mathcal{L}_{\overline{\phi}}^{n} (\mathbb{1})(x)} - \frac{\frac{1}{n} \sum_{j=0}^{n-1} u_{\phi}(x) \int u \mathcal{L}_{\overline{\phi}}^{j} (\mathbb{1}) \, \mathrm{d} m_{\phi}}{u_{\phi}(x)} \right| = 0,$$

uniformly in $u \in C_h^b(S^2, d)$, $\phi \in H$, and $x \in S^2$. Combining the above with (5.69), (5.70), (5.72), and the calculation in the beginning part of the proof, we can conclude, therefore, that the left-hand side of (5.68) is equal to

$$\lim_{n \to +\infty} \frac{1}{n} \sum_{j=0}^{n-1} \int u \mathcal{L}_{\overline{\phi}}^{j} (\mathbb{1}) \, \mathrm{d} m_{\phi} = \lim_{n \to +\infty} \frac{1}{n} \sum_{j=0}^{n-1} \int u u_{\phi} \, \mathrm{d} m_{\phi} = \int u \, \mathrm{d} \mu_{\phi},$$

and the convergence is uniform in $u \in C_h^b(S^2, d)$ and $\phi \in H$. □

We record the following well-known fact for the convenience of the reader.

Lemma 5.34 *For each metric d on S^2 that generates the standard topology on S^2 and each $\alpha \in (0, 1]$, $C^{0,\alpha}(S^2, d)$ is a dense subset of $C(S^2)$ with respect to the uniform norm. In particular, $C^{0,\alpha}(S^2, d)$ is a dense subset of $C(S^2)$ in the weak topology.*

Proof The lemma follows from the fact that the set of Lipschitz functions are dense in $C(S^2)$ with respect to the uniform norm (see for example, [He01, Theorem 6.8]). □

Theorem 5.35 *Let $f : S^2 \to S^2$ be an expanding Thurston map, and d be a visual metric on S^2 for f. Let $\phi, \gamma \in C^{0,\alpha}(S^2, d)$ be real-valued Hölder continuous functions with an exponent $\alpha \in (0, 1]$. Then for each $t \in \mathbb{R}$, we have*

$$\frac{\mathrm{d}}{\mathrm{d}t} P(f, \phi + t\gamma) = \int \gamma \, \mathrm{d} \mu_{\phi + t\gamma}. \tag{5.75}$$

Proof We will use the well-known fact from real analysis that if a sequence $\{g_n\}_{n \in \mathbb{N}}$ of real-valued differentiable functions defined on a finite interval in $\mathbb{R}$ converges pointwise to some function g and the sequence of the corresponding derivatives $\left\{ \frac{\mathrm{d} g_n}{\mathrm{d}t} \right\}_{n \in \mathbb{N}}$ converges uniformly to some function h, then g is differentiable and $\frac{\mathrm{d}g}{\mathrm{d}t} = h$.

Fix a point $x \in S^2$ and a constant $l \in (0, +\infty)$. For $n \in \mathbb{N}$ and $t \in \mathbb{R}$, define

$$P_n(t) = \frac{1}{n} \log \sum_{y \in f^{-n}(x)} \deg_{f^n}(y) \exp(S_n(\phi + t\gamma)(y)). \tag{5.76}$$

Observe that there exists a bounded subset H of $C^{0,\alpha}(S^2, d)$ such that $\phi + t\gamma \in H$ for each $t \in (-l, l)$. Then by Lemma 5.33,

$$\frac{\mathrm{d}P_n}{\mathrm{d}t}(t) = \frac{\frac{1}{n} \sum\limits_{y \in f^{-n}(x)} \deg_{f^n}(y)(S_n\gamma(y)) \exp(S_n(\phi + t\gamma)(y))}{\sum\limits_{y \in f^{-n}(x)} \deg_{f^n}(y) \exp(S_n(\phi + t\gamma)(y))} \tag{5.77}$$

converges to $\int \gamma \,\mathrm{d}\mu_{\phi+t\gamma}$ as $n \longrightarrow +\infty$, uniformly in $t \in (-l, l)$.

On the other hand, by Proposition 5.19, for each $t \in (-l, l)$, we have

$$\lim_{n \to +\infty} P_n(t) = P(f, \phi + t\gamma). \tag{5.78}$$

Hence $P(f, \phi + t\gamma)$ is differentiable with respect to t on $(-l, l)$ and

$$\frac{\mathrm{d}}{\mathrm{d}t} P(f, \phi + t\gamma) = \lim_{n \to +\infty} \frac{\mathrm{d}P_n}{\mathrm{d}t}(t) = \int \gamma \,\mathrm{d}\mu_{\phi+t\gamma}.$$

Since $l \in (0, +\infty)$ is arbitrary, the proof is complete. □

Theorem 5.36 *Let $f: S^2 \to S^2$ be an expanding Thurston map and d be a visual metric on S^2 for f. Let $\phi \in C^{0,\alpha}(S^2, d)$ be a real-valued Hölder continuous function with an exponent $\alpha \in (0, 1]$. Then there exists a unique equilibrium state μ_ϕ for f and ϕ. Moreover, the map f with respect to μ_ϕ is forward quasi-invariant (i.e., for each Borel set $A \subseteq S^2$, if $\mu_\phi(A) = 0$, then $\mu_\phi(f(A)) = 0$), and nonsingular (i.e., for each Borel set $A \subseteq S^2$, $\mu_\phi(A) = 0$ if and only if $\mu_\phi(f^{-1}(A)) = 0$).*

Proof The existence is proved in Corollary 5.20.

We now prove the uniqueness.

Since $\phi \in C^{0,\alpha}(S^2, d)$, by Theorem 5.35 the function

$$t \longmapsto P(f, \phi + t\gamma)$$

is differentiable at 0 for $\gamma \in C^{0,\alpha}(S^2, d)$. Recall that by Lemma 5.34 $C^{0,\alpha}(S^2, d)$ is dense in $C(S^2)$ in the weak topology. We note that the topological pressure function $P(f, \cdot): C(S^2) \to \mathbb{R}$ is convex continuous (see for example, [PU10, Theorem 3.6.1 and Theorem 3.6.2]). Thus by Theorem 5.23 with $V = C(S^2)$, $x = \phi$, $U = C^{0,\alpha}(S^2, d)$, and $Q = P(f, \cdot)$, we get $\operatorname{card}\left(V^*_{\phi, P(f,\cdot)}\right) = 1$.

On the other hand, if μ is an equilibrium state for f and ϕ, then by (3.4) and (3.5),

$$h_\mu(f) + \int \phi \,\mathrm{d}\mu = P(f, \phi),$$

and for each $\gamma \in C(S^2)$,

$$h_\mu(f) + \int (\phi + \gamma) \,\mathrm{d}\mu \leq P(f, \phi + \gamma).$$

So $\int \gamma \,\mathrm{d}\mu \leq P(f, \phi + \gamma) - P(f, \phi)$. Thus by (5.44), the continuous functional $\gamma \longmapsto \int \gamma \,\mathrm{d}\mu$ on $C(S^2)$ is in $V^*_{\phi, P(f,\cdot)}$. Since $\mu_\phi = u_\phi m_\phi$ defined in Theorem 5.17 is an equilibrium state for f and ϕ, and $\operatorname{card}\left(V^*_{\phi,P(f,\cdot)}\right) = 1$, we get that each equilibrium state μ for f and ϕ must satisfy $\int \gamma \,\mathrm{d}\mu = \int \gamma \,\mathrm{d}\mu_\phi$ for $\gamma \in \mathscr{C}(S^2)$, i.e., $\mu = \mu_\phi$.

The fact that the map f is forward quasi-invariant and nonsingular with respect to μ_ϕ follows from the corresponding result for m_ϕ in Theorem 5.12, Lemma 5.26, and the fact that $m_{\widetilde{\phi}} = \mu_\phi$ from Corollary 5.32. □

Remark 5.37 Since the entropy map $\mu \longmapsto h_\mu(f)$ for an expanding Thurston map f is *affine* (see for example, [Wal82, Theorem 8.1]), i.e., if $\mu, \nu \in \mathscr{M}(S^2, f)$ and $p \in [0, 1]$, then $h_{p\mu+(1-p)\nu}(f) = p h_\mu(f) + (1 - p) h_\nu(f)$, so is the pressure map $\mu \longmapsto P_\mu(f, \phi)$ for f and a Hölder continuous potential $\phi\colon S^2 \to \mathbb{R}$. Thus the uniqueness of the equilibrium state μ_ϕ and the Variational Principle (3.5) imply that μ_ϕ is an extreme point of the convex set $\mathscr{M}(S^2, f)$. It follows from the fact (see for example, [PU10, Theorem 2.2.8]) that the extreme points of $\mathscr{M}(S^2, f)$ are exactly the ergodic measures in $\mathscr{M}(S^2, f)$ that μ_ϕ is ergodic. However, we are going to prove a much stronger ergodic property of μ_ϕ in Sect. 5.4.

The following proposition is an immediate consequence of Theorem 5.31.

Proposition 5.38 *Let f, d, ϕ satisfy the Assumptions. Let μ_ϕ be the unique equilibrium state for f and ϕ. Then for each Borel probability measure $\mu \in \mathscr{P}(S^2)$, we have*

$$\left(\mathscr{L}^*_{\widetilde{\phi}}\right)^n(\mu) \xrightarrow{w^*} \mu_\phi \text{ as } n \longrightarrow +\infty. \tag{5.79}$$

Proof Recall that for each $u \in C(S^2)$, there exists some abstract modulus of continuity h such that $u \in C^\beta_h(S^2, d)$, where $\beta = \|u\|_\infty$. By Theorem 5.36 and Theorem 5.17, we have $\mu_\phi = u_\phi m_\phi$ as constructed in Theorem 5.17. Then by Lemma 5.25 and (5.60) in Theorem 5.31,

$$\begin{aligned}\lim_{n\to+\infty} \left\langle \left(\mathscr{L}^*_{\widetilde{\phi}}\right)^n(\mu), u \right\rangle &= \lim_{n\to+\infty} \left(\left\langle \mu, \mathscr{L}^n_{\widetilde{\phi}}(u - \langle \mu_\phi, u\rangle \mathbb{1}) \right\rangle + \left\langle \mu, \mathscr{L}^n_{\widetilde{\phi}}(\langle \mu_\phi, u\rangle \mathbb{1}) \right\rangle \right) \\ &= 0 + \langle \mu, \langle \mu_\phi, u\rangle \mathbb{1}\rangle = \langle \mu_\phi, u\rangle,\end{aligned}$$

for each $u \in C(S^2)$. Therefore, (5.79) holds. □

5.4 Ergodic Properties

In this section, we first prove that if $f, \mathscr{C}, d$, and ϕ satisfies the Assumptions, then any edge in the cell decompositions induced by f and $\mathscr{C}$ is a zero set with respect to the measures m_ϕ or μ_ϕ. This result is also important for Theorem 5.45. We then show in Theorem 5.41 that the measure-preserving transformation f of the probability space (S^2, μ_ϕ) is *exact* (Definition 5.40), and as an immediate consequence, mixing (Corollary 5.44). Another consequence of Theorem 5.41 is that μ_ϕ is non-atomic (Corollary 5.42).

Proposition 5.39 *Let $f, \mathscr{C}, n_\mathscr{C}, d, \phi, \alpha$ satisfy the Assumptions. Let μ_ϕ be the unique equilibrium state for f and ϕ, and m_ϕ be as in Corollary 5.32. Then*

$$m_\phi\left(\bigcup_{i=0}^{+\infty} f^{-i}(\mathscr{C})\right) = \mu_\phi\left(\bigcup_{i=0}^{+\infty} f^{-i}(\mathscr{C})\right) = 0. \tag{5.80}$$

Proof Since $\mu_\phi \in \mathscr{M}(S^2, f)$ is f-invariant, and $\mathscr{C} \subseteq f^{-in_\mathscr{C}}(\mathscr{C})$ for each $i \in \mathbb{N}$, we have $\mu_\phi\left(f^{-in_\mathscr{C}}(\mathscr{C}) \setminus \mathscr{C}\right) = 0$ for each $i \in \mathbb{N}$. Since f is expanding, by Lemma 5.11, there exist $m \in \mathbb{N}$ and an $(mn_\mathscr{C})$-tile $X \in \mathbf{X}^{mn_\mathscr{C}}$ such that $X \cap \mathscr{C} = \emptyset$. Then $\partial X \subseteq f^{mn_\mathscr{C}}(\mathscr{C}) \setminus \mathscr{C}$. So $\mu_\phi(\partial X) = 0$. Since $\mu_\phi = u_\phi m_\phi$, where u_ϕ is bounded away from 0 (see Theorem 5.17), we have $m_\phi(\partial X) = 0$. Note that $f^{mn_\mathscr{C}}|_{\partial X}$ is a homeomorphism from ∂X to $\mathscr{C}$ (see Proposition 2.6). Thus by the information on the Jacobian for f with respect to m_ϕ in Theorem 5.12, we get $m_\phi(\mathscr{C}) = 0$.

Now suppose there exist $k \in \mathbb{N}$ and a k-edge $e \in \mathbf{E}^k$ such that $m_\phi(e) > 0$. Then by using the Jacobian for f with respect to m_ϕ again, we get $m_\phi(\mathscr{C}) > 0$, a contradiction. Hence $m_\phi\left(\bigcup_{i=0}^{+\infty} f^{-i}(\mathscr{C})\right) = 0$. Since $\mu_\phi = u_\phi m_\phi$, we get $\mu_\phi\left(\bigcup_{i=0}^{+\infty} f^{-i}(\mathscr{C})\right) = 0$. □

For each Borel measure μ on a compact metric space (X, d), we denote by $\overline{\mu}$ the *completion* of μ, i.e., $\overline{\mu}$ is the unique measure defined on the smallest σ-algebra $\overline{\mathscr{B}}$ containing all Borel sets and all subsets of μ-null sets, satisfying $\overline{\mu}(E) = \mu(E)$ for each Borel set $E \subseteq X$.

Definition 5.40 Let g be a measure-preserving transformation of a probability space (X, μ). Then g is called *exact* if for every measurable set E with $\mu(E) > 0$ and measurable images $g(E), g^2(E), \dots$, the following holds:

$$\lim_{n\to+\infty} \mu\left(g^n(E)\right) = 1.$$

Note that in Definition 5.40, we do not require μ to be a Borel measure. In the case when g is a Thurston map on S^2 and μ is a Borel measure, the set $g^n(E)$ is a Borel set for each $n \in \mathbb{N}$ and each Borel set $E \subseteq S^2$. Indeed, a Borel set $E \subseteq S^2$ can be covered by n-tiles in the cell decompositions of S^2 induced by g and any Jordan curve $\mathscr{C} \subseteq S^2$ containing post g. For each n-tile $X \in \mathbf{X}^n(f, \mathscr{C})$, the restriction $g^n|_X$

of g^n to X is a homeomorphism from the closed set X onto $g^n(X)$ by Proposition 2.6. It is then clear that the set $g^n(E)$ is also Borel.

We now prove that the measure-preserving transformation f of the probability space (S^2, μ_ϕ) is exact. The argument that we use here is similar to that in the proof of the exactness of an open, topologically exact, distance-expanding self-map of a compact metric space equipped with a certain Gibbs state ([PU10, Theorem 5.2.12]).

Theorem 5.41 *Let $f: S^2 \to S^2$ be an expanding Thurston map and d be a visual metric on S^2 for f. Let $\phi \in C^{0,\alpha}(S^2, d)$ be a real-valued Hölder continuous function with an exponent $\alpha \in (0, 1]$. Let μ_ϕ be the unique equilibrium state for f and ϕ, and $\overline{\mu_\phi}$ its completion.*

Then the measure-preserving transformation f of the probability space (S^2, μ_ϕ) (resp. $(S^2, \overline{\mu_\phi})$) is exact.

Proof We fix a Jordan curve $\mathscr{C} \subseteq S^2$ that satisfies the Assumptions (see Theorem 2.16 for the existence of such $\mathscr{C}$).

Since $\mu_\phi = u_\phi m_\phi$, by (5.35), it suffices to prove that

$$\lim_{n\to+\infty} m_\phi(S^2 \setminus f^n(A)) = 0$$

for each Borel set $A \subseteq S^2$ with $m_\phi(A) > 0$.

Let $A \subseteq S^2$ be an arbitrary Borel subset of S^2 with $m_\phi(A) > 0$. Then there exists a compact set $E \subseteq A$ such that $m_\phi(E) > 0$. Fix an arbitrary $\varepsilon > 0$. Since f is expanding, by Lemma 5.11, n-tiles have uniformly small diameters if n is large. This and the outer regularity of the Borel measures enable us to choose $N \in \mathbb{N}$ such that for each $n \geq N$, the collection

$$\mathbf{P}^n = \{X^n \in \mathbf{X}^n(f, \mathscr{C}) \mid X^n \cap E \neq \emptyset\}$$

of n-tiles satisfies $m_\phi\left(\bigcup \mathbf{P}^n\right) \leq m_\phi(E) + \varepsilon$. Thus for each $n \geq N$, we have $m_\phi\left(\bigcup\limits_{X^n\in\mathbf{P}^n} X^n \setminus E\right) \leq \varepsilon$. So $\sum\limits_{X^n\in\mathbf{P}^n} m_\phi(X^n \setminus E) \leq \varepsilon$ by Proposition 5.39. Hence

$$\frac{\sum\limits_{X^n\in\mathbf{P}^n} m_\phi(X^n \setminus E)}{\sum\limits_{X^n\in\mathbf{P}^n} m_\phi(X^n)} \leq \frac{\varepsilon}{m_\phi(E)}. \tag{5.81}$$

Thus for each $n \geq N$, there exists some n-tile $Y^n \in \mathbf{P}^n$ such that

$$\frac{m_\phi(Y^n \setminus E)}{m_\phi(Y^n)} \leq \frac{\varepsilon}{m_\phi(E)}. \tag{5.82}$$

By Proposition 2.6(i), the map f^n is injective on Y^n. So by Theorem 5.12, Lemma 5.3, (5.7), and (5.82), we have

$$
\begin{aligned}
\frac{m_\phi\left(f^n(Y^n)\setminus f^n(E)\right)}{m_\phi\left(f^n(Y^n)\right)} &\leq \frac{m_\phi\left(f^n(Y^n\setminus E)\right)}{m_\phi\left(f^n(Y^n)\right)} \\
&= \frac{\int_{Y^n\setminus E}\exp(-S_n\phi)\,\mathrm{d}m_\phi}{\int_{Y^n}\exp(-S_n\phi)\,\mathrm{d}m_\phi} \leq C_2^2\frac{m_\phi(Y^n\setminus E)}{m_\phi(Y^n)} \leq \frac{C_2^2\varepsilon}{m_\phi(E)},
\end{aligned}
$$

where $C_2 \geq 1$ is the constant defined in (5.7) that depends only on $f, \mathscr{C}, d, \phi$, and α. By Lemma 5.11, there exists $k \in \mathbb{N}$ that depends only on f and $\mathscr{C}$ such that $f^k(X_w^0) = f^k(X_b^0) = S^2$, where X_w^0 and X_b^0 are the white 0-tile and the black 0-tile, respectively. Since $f^n(Y^n)$ is either X_w^0 or X_b^0, by Proposition 5.13, for each $n \geq N$,

$$
\begin{aligned}
& m_\phi\left(S^2\setminus f^{n+k}(E)\right) \leq m_\phi\left(f^k\left(f^n(Y^n)\setminus f^n(E)\right)\right) \\
\leq & \int_{f^n(Y^n)\setminus f^n(E)}\exp(-S_k\phi)\,\mathrm{d}m_\phi \leq \exp(k\left\|\phi\right\|_\infty)\frac{C_2^2\varepsilon}{m_\phi(E)}.
\end{aligned}
$$

Since $\varepsilon > 0$ was arbitrary, we get

$$
\lim_{n\to+\infty} m_\phi\left(S^2\setminus f^{n+k}(E)\right) = 0. \tag{5.83}
$$

Thus

$$
\lim_{n\to+\infty} m_\phi\left(f^n(A)\right) \geq \lim_{n\to+\infty} m_\phi\left(f^n(E)\right) = 1.
$$

Hence the measure-preserving transformation f of the probability space (S^2, μ_ϕ) is exact.

Next, we observe that since f is μ_ϕ-measurable, and is a measure-preserving transformation of the probability space (S^2, μ_ϕ), it is clear that f is also $\overline{\mu_\phi}$-measurable, and is a measure-preserving transformation of the probability space $(S^2, \overline{\mu_\phi})$.

To prove that the measure-preserving transformation f of the probability space $(S^2, \overline{\mu_\phi})$ is exact, we consider a $\overline{\mu_\phi}$-measurable set $B \subseteq S^2$ with $\overline{\mu_\phi}(B) > 0$. Since $\overline{\mu_\phi}$ is the completion of the Borel probability measure μ_ϕ, we can choose Borel sets A and C such that $A \subseteq B \subseteq C \subseteq S^2$ and $\overline{\mu_\phi}(B) = \overline{\mu_\phi}(A) = \overline{\mu_\phi}(C) = \mu_\phi(A) = \mu_\phi(C)$. For each $n \in \mathbb{N}$, we have $f^n(A) \subseteq f^n(B) \subseteq f^n(C)$ and both $f^n(A)$ and $f^n(C)$ are Borel sets (see the discussion following Definition 5.40). Since f is forward quasi-invariant with respect to μ_ϕ (see Theorem 5.36), it is clear that $\mu_\phi(f^n(A)) = \mu_\phi(f^n(C))$. Thus

$$
\mu_\phi\left(f^n(A)\right) = \overline{\mu_\phi}\left(f^n(A)\right) = \overline{\mu_\phi}\left(f^n(B)\right) = \overline{\mu_\phi}\left(f^n(C)\right) = \mu_\phi\left(f^n(C)\right).
$$

Therefore, $\lim_{n\to+\infty} \overline{\mu_\phi}(f^n(B)) = \lim_{n\to+\infty} \mu_\phi(f^n(A)) = 1.$ □

Let μ be a measure on a topological space X. Then μ is called *non-atomic* if $\mu(\{x\}) = 0$ for each $x \in X$.

The following corollary strengthens Theorem 5.12.

Corollary 5.42 *Let f, d, ϕ, α satisfy the Assumptions. Let μ_ϕ be the unique equilibrium state for f and ϕ, and m_ϕ be as in Corollary 5.32. Then both μ_ϕ and m_ϕ as well as their corresponding completions are non-atomic.*

Proof Since $\mu_\phi = u_\phi m_\phi$, where u_ϕ is bounded away from 0 (see Theorem 5.17), it suffices to prove that μ_ϕ is non-atomic.

Suppose there exists a point $x \in S^2$ with $\mu_\phi(\{x\}) > 0$, then for all $y \in S^2$, we have

$$\mu_\phi(\{y\}) \leq \max\{\mu_\phi(\{x\}), 1 - \mu_\phi(\{x\})\}.$$

Since the transformation f of (S^2, μ_ϕ) is exact by Theorem 5.41, we get that $\mu_\phi(\{x\}) = 1$ and $f(x) = x$.

We fix a Jordan curve $\mathscr{C} \subseteq S^2$ that satisfies the Assumptions (see Theorem 2.16 for the existence of such $\mathscr{C}$). It is clear from Lemma 5.11 that there exist $n \in \mathbb{N}$ and an n-tile $X^n \in \mathbf{X}^n(f, \mathscr{C})$ with $x \notin X^n$. Then $\mu_\phi(X^n) = 0$, which contradicts with the fact that μ_ϕ is a Gibbs state for f, $\mathscr{C}$, and ϕ (see Theorem 5.17 and Definition 5.5).

The fact that the completions are non-atomic now follows immediately. □

Let f, d, ϕ, α satisfy the Assumptions. Let μ_ϕ be the unique equilibrium state for f and ϕ, and $\overline{\mu_\phi}$ its completion. Then by Theorem 2.7 in [Ro49], the complete separable metric space (S^2, d) equipped the complete non-atomic measure $\overline{\mu_\phi}$ is a Lebesgue space in the sense of V. Rokhlin. We omit V. Rokhlin's definition of a *Lebesgue space* here and refer the reader to [Ro49, Sect. 2], since the only results we will use about Lebesgue spaces are V. Rokhlin's definition of exactness of a measure-preserving transformation on a Lebesgue space and its implication to the mixing properties. More precisely, in [Ro61], V. Rokhlin gave a definition of exactness for a measure-preserving transformation on a Lebesgue space equipped with a complete non-atomic measure, and showed [Ro61, Sect. 2.2] that in such a context, it is equivalent to our definition of exactness in Definition 5.40. Moreover, he proved [Ro61, Sect. 2.6] that if a measure-preserving transformation on a Lebesgue space equipped with a complete non-atomic measure is exact, then it is *mixing* (he actually proved that it is *mixing of all degrees*, which we will not discuss here).

Let us recall the definition of mixing for a measure-preserving transformation.

Definition 5.43 Let g be a measure-preserving transformation of a probability space (X, μ). Then g is called *mixing* if for all measurable sets $A, B \subseteq X$, the following holds:

$$\lim_{n \to +\infty} \mu\left(g^{-n}(A) \cap B\right) = \mu(A) \cdot \mu(B).$$

We call g *ergodic* if for each measurable set $E \subseteq X$, $g^{-1}(E) = E$ implies either $\mu(E) = 0$ or $\mu(E) = 1$.

It is well-known and easy to see that if g is mixing, then it is ergodic (see for example, [Wal82]).

Corollary 5.44 *Let f, d, ϕ, α satisfy the Assumptions. Let μ_ϕ be the unique equilibrium state for f and ϕ, and $\overline{\mu_\phi}$ its completion. Then the measure-preserving transformation f of the probability space (S^2, μ_ϕ) (resp. $(S^2, \overline{\mu_\phi})$) is mixing and ergodic.*

Proof By the discussion preceding Definition 5.43, we know that the measure-preserving transformation f of $(S^2, \overline{\mu_\phi})$ is mixing and thus ergodic. Since any μ_ϕ-measurable sets $A, B \subseteq S^2$ are also $\overline{\mu_\phi}$-measurable, the measure-preserving transformation f of (S^2, μ_ϕ) is also mixing and ergodic. □

5.5 Co-homologous Potentials

The goal of this section is to prove in Theorem 5.45 that two equilibrium states are identical if and only if there exists a constant $K \in \mathbb{R}$ such that $K\mathbb{1}_{S^2}$ and the difference of the corresponding potentials are *co-homologous* (see Definition 5.46). We use some of the ideas from [PU10] in the process of proving Theorem 5.45. We establish a form of the *closing lemma* for expanding Thurston maps in Lemma 5.50.

Theorem 5.45 *Let $f : S^2 \to S^2$ be an expanding Thurston map, and d be a visual metric on S^2 for f. Let $\phi, \psi \in C^{0,\alpha}(S^2, d)$ be real-valued Hölder continuous functions with an exponent $\alpha \in (0, 1]$. Let μ_ϕ (resp. μ_ψ) be the unique equilibrium state for f and ϕ (resp. ψ). Then $\mu_\phi = \mu_\psi$ if and only if there exists a constant $K \in \mathbb{R}$ such that $\phi - \psi$ and $K\mathbb{1}_{S^2}$ are co-homologous in the space $C(S^2)$ of real-valued continuous functions.*

Definition 5.46 Let $g : X \to X$ be a continuous map on a metric space (X, d). Let $\mathscr{K} \subseteq C(X, \mathbb{C})$ be a subspace of the space $C(X, \mathbb{C})$ of complex-valued continuous function on X. Two functions $\phi, \psi \in C(X, \mathbb{C})$ are said to be *co-homologous (in $\mathscr{K}$)* if there exists $u \in \mathscr{K}$ such that $\phi - \psi = u \circ g - u$.

Remark 5.47 As we will see in the proof of Theorem 5.45 at the end of this section, if $\mu_\phi = \mu_\psi$ then the corresponding u can be chosen from $C^{0,\alpha}(S^2, d)$.

Lemma 5.48 *Let f and $\mathscr{C}$ satisfy the Assumptions. If $f(\mathscr{C}) \subseteq \mathscr{C}$, then for $m, n \in \mathbb{N}$ with $m \geq n$ and each m-vertex $v^m \in \mathbf{V}^m(f, \mathscr{C})$ with $\overline{W}^m(v^m) \subseteq W^{m-n}(f^n(v^m))$, there exists $x \in \overline{W}^m(v^m)$ such that $f^n(x) = x$.*

Here $\overline{W}^m(v^m)$ denotes the closure of the open set $W^m(v^m)$.

Proof Since $v^m \in W^{m-n}(f^n(v^m))$ and $f(\mathscr{C}) \subseteq \mathscr{C}$, depending on the location of v^m, there are exactly three cases, namely, (i) $v^m = f^n(v^m)$; (ii) v^m is contained in the interior of some $(m-n)$-edge; (iii) v^m is contained in the interior of some $(m-n)$-tile. We will find a fixed point $x \in \overline{W}^m(v^m)$ of f^n in each case.

Case 1. When $v^m = f^n(v^m)$, we can just set $x = v^m$.

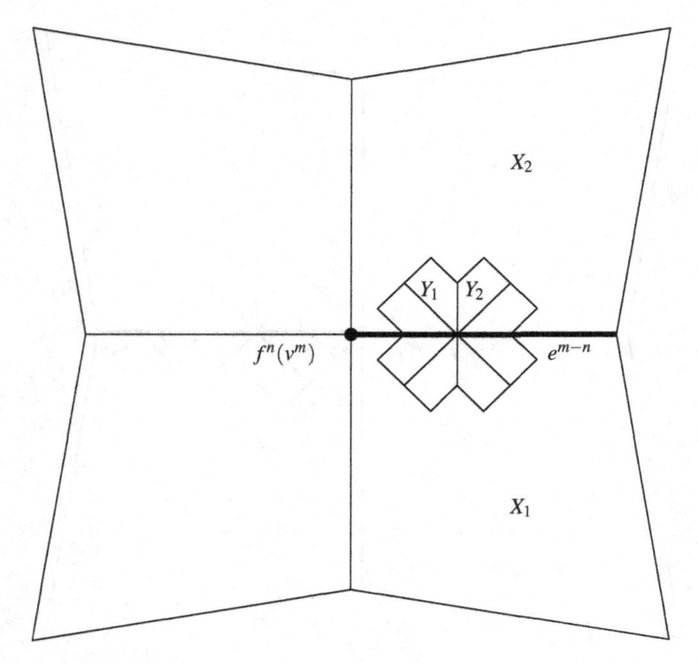

Fig. 5.2 An example for Case 2 when $Y_2 \subseteq X_2$

Case 2. When $v^m \in \mathrm{inte}(e^{m-n})$ for some $(m-n)$-edge $e^{m-n} \in \mathbf{E}^{m-n}$ with $\mathrm{inte}\,(e^{m-n}) \subseteq W^{m-n}\,(f^n(v^m))$, it is clear that $W^m(v^m) \subseteq X_1 \cup X_2$ when $X_1, X_2 \in \mathbf{X}^{m-n}$ form the unique pair of distinct $(m-n)$-tiles contained in $W^{m-n}\,(f^n(v^m))$ with $X_1 \cap X_2 = e^{m-n}$. We can choose a pair of distinct m-tiles $Y_1, Y_2 \in \mathbf{X}^m$ with $Y_1 \cup Y_2 \subseteq \overline{W}^m(v^m)$, $f^n(Y_1) = X_1$, $f^n(Y_2) = X_2$, and $Y_1 \cap Y_2 = e^m$ for some m-edge $e^m \in \mathbf{E}^m$. If either $Y_1 \subseteq X_1$ or $Y_2 \subseteq X_2$, say $Y_2 \subseteq X_2$, then since X_2 is homeomorphic to the closed unit disk in $\mathbb{R}^2$, and f^n maps Y_2 homeomorphically onto X_2 (Proposition 2.6(i)), we can conclude by applying Brouwer's Fixed Point Theorem on $((f^n)|_{Y_2})^{-1}$ that there exists a fixed point $x \in Y_2$ of f^n. (See for example, Fig. 5.2.) So we can assume without loss of generality that $Y_1 \subseteq X_2$ and $Y_2 \subseteq X_1$. Suppose now that $\mathrm{inte}(e^m) \subseteq \mathrm{inte}(X_i)$, then $Y_1 \cup Y_2 \subseteq X_i$, for $i \in \{1, 2\}$. So $e^m \subseteq e^{m-n}$. Since f^n maps e^m homeomorphically onto e^{m-n} by Proposition 2.6(i), and e^{m-n} is homeomorphic to the closed unit interval in $\mathbb{R}$, it is clear that there exists a fixed point $x \in e^m$ of f^n. (See for example, Fig. 5.3.)

Case 3. When $v^m \in \mathrm{inte}(X^{m-n})$ for some $(m-n)$-tile $X^{m-n} \in \mathbf{X}^{m-n}$ contained in $\overline{W}^{m-n}(f^n(v^m))$, it is clear that $W^m(v^m) \subseteq X^{m-n}$. Let $X^m \in \mathbf{X}^m$ be an m-tile contained in $\overline{W}^m(v^m)$ such that $f^n(X^m) = X^{m-n}$. Since X^{m-n} is homeomorphic to the closed unit disk in $\mathbb{R}^2$, and f^n maps X^m homeomorphically onto X^{m-n} (Proposition 2.6(i)), we can conclude by applying Brouwer's Fixed Point Theorem on $((f^n)|_{X^m})^{-1}$ that there exists a fixed point $x \in X^m$ of f^n. $\square$

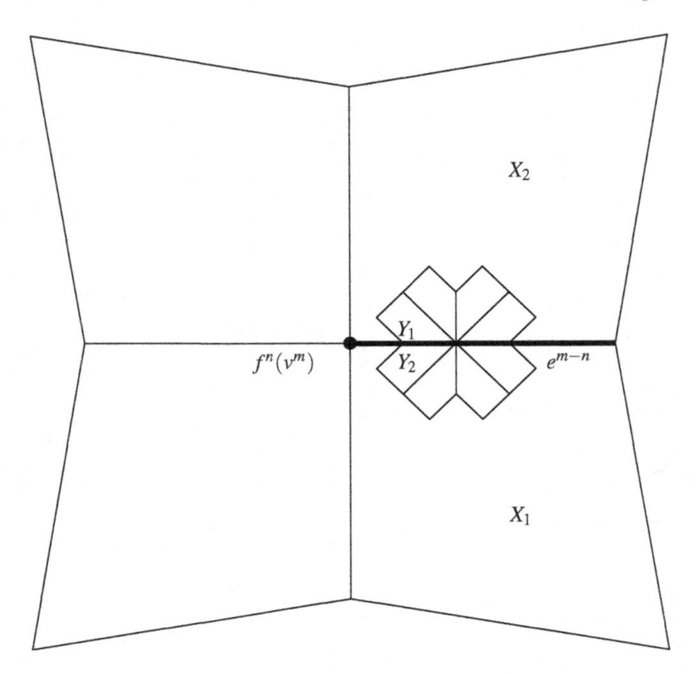

Fig. 5.3 An example for Case 2 when $Y_1 \not\subseteq X_1$, $Y_2 \not\subseteq X_2$

Lemma 5.49 *Let f and $\mathscr{C}$ satisfy the Assumptions. Then there exists a number $\kappa \in \mathbb{N}_0$ such that the following statement holds:*

For each $x \in S^2$, each $n \in \mathbb{N}_0$, and each n-tile $X^n \in \mathbf{X}^n(f, \mathscr{C})$, if $x \in X^n$, then there exists an n-vertex $v^n \in \mathbf{V}^n(f, \mathscr{C}) \cap X^n$ with

$$U^{n+\kappa}(x) \subseteq W^n(v^n). \tag{5.84}$$

Proof We will first find $\kappa \in \mathbb{N}_0$ such that the statement above holds when $n = 0$. We will then show that the same κ works for arbitrary $n \in \mathbb{N}_0$.

We fix a visual metric d on S^2 for f with expansion factor $\Lambda > 1$.

Note that the collection of 0-flowers $\{W^0(v^0) \mid v^0 \in \mathbf{V}^0\}$ forms a finite open cover of S^2. By the Lebesgue Number Lemma ([Mu00, Lemma 27.5]), there exists a number $\varepsilon > 0$ such that any set of diameter at most ε is a subset of $W^0(v^0)$ for some $v^0 \in \mathbf{V}^0$. Here ε depends only on f, $\mathscr{C}$, and d. Then by Proposition 2.13(iii), there exists $\kappa \in \mathbb{N}_0$ depending only on $f, \mathscr{C}$, and d such that $\operatorname{diam}_d(U^\kappa(x)) < \varepsilon$ for $x \in S^2$. So for each $x \in S^2$, there exists a 0-vertex $v^0 \in \mathbf{V}^0$ such that $U^\kappa(x) \subseteq W^0(v^0)$. Let $X^0 \in \mathbf{X}^0$ be a 0-tile with $x \in X^0$, then clearly $v^0 \in X^0$.

In general, we fix $x \in S^2$, $n \in \mathbb{N}_0$, and $X^n \in \mathbf{X}^n$ with $x \in X^n$. Set $A = \mathbf{V}^n \cap X^n$. By Proposition 2.6, we have $f^n(W^n(v^n)) = W^0(f^n(v^n))$ and $f^n(\partial W^n(v^n)) = \partial W^0(f^n(v^n))$ for each $v^n \in A$. Suppose $U^{n+\kappa}(x) \not\subseteq W^n(v^n)$ for all $v^n \in A$. Since

$x \in X^n$ and $U^{n+\kappa}(x)$ is connected, we have $U^{n+\kappa}(x) \cap \partial W^n(v^n) \neq \emptyset$, and thus by Proposition 2.6(i)

$$U^\kappa(f^n(x)) \cap \partial W^0(f^n(v^n)) \supseteq f^n(U^{n+\kappa}(x)) \cap f^n(\partial W^n(v^n)) \neq \emptyset,$$

for each $v^n \in A$. Since $f^n(A) = \mathbf{V}^0$ by Proposition 2.6, it follows that $U^\kappa(f^n(x)) \not\subseteq W^0(v^0)$ for all $v^0 \in \mathbf{V}^0$, contradicting the discussion above for the case when $n = 0$.

Finally, we note that (5.84) holds or fails independently of the choice of d. Therefore, the number κ depends only on f and $\mathcal{C}$. □

The following result can be considered as a form of the *closing lemma* for expanding Thurston maps. It is a key ingredient in the proof of Proposition 5.52, which will be used to prove Theorem 5.45. Note that Lemma 5.50 is more technical and in some sense slightly stronger than the closing lemma for forward-expansive maps (see [PU10, Corollary 4.2.5]). We need it in this slightly stronger form, since the distortion lemmas (Lemmas 5.3 and 5.4) cannot be applied in the proof of Proposition 5.52.

Lemma 5.50 (Closing lemma) *Let $f, \mathcal{C}, d, \Lambda$ satisfy the Assumptions. If $f(\mathcal{C}) \subseteq \mathcal{C}$, then there exist $M \in \mathbb{N}_0$, $\delta_0 \in (0, 1)$, and $\beta_0 > 1$ such that the following statement holds:*

For each $\delta \in (0, \delta_0]$, if $x \in S^2$ and $l \in \mathbb{N}$ satisfy $l > M$ and $d(x, f^l(x)) \leq \delta$, then there exists $y \in S^2$ such that $f^l(y) = y \in U^{N+l}(x)$ and $d(f^i(x), f^i(y)) \leq \beta_0 \delta \Lambda^{-(l-i)}$ for each $i \in \{0, 1, \dots, l\}$, where $N = \lceil -\log_\Lambda(\delta_0^{-1}\delta) \rceil \in \mathbb{N}_0$.

Proof Define

$$\delta_0 = (2K)^{-1}\Lambda^{-(\kappa+1)}, \tag{5.85}$$

$$\beta_0 = 4K^2\Lambda^{\kappa+1} = 2K\delta_0^{-1}, \tag{5.86}$$

$$M = \lceil \log_\Lambda(10K^2) \rceil + \kappa \in \mathbb{N}_0, \tag{5.87}$$

where $K \geq 1$ and $\kappa \in \mathbb{N}_0$ are constants depending only on f, $\mathcal{C}$, and d from Lemma 2.13 and Lemma 5.49, respectively.

We fix $\delta \in (0, \delta_0]$ and set

$$\beta = \beta_0\delta. \tag{5.88}$$

Note that $N = \lceil -\log_\Lambda(\delta_0^{-1}\delta) \rceil = \lceil -\log_\Lambda \frac{\beta}{2K} \rceil \in \mathbb{N}_0$ by (5.88) and (5.86). So

$$2K\Lambda^{-N} \leq \beta \leq 2K\Lambda^{-N+1}, \tag{5.89}$$

and by (5.88) and (5.86), we have

$$\delta \leq (2K)^{-1}\Lambda^{-(N+\kappa)}. \tag{5.90}$$

Recall that by Lemma 2.13(iii), for $z \in S^2$ and $n \in \mathbb{N}_0$, we have

$$B_d(z, K^{-1}\Lambda^{-n}) \leq U^n(z) \leq B_d(z, K\Lambda^{-n}). \tag{5.91}$$

Fix $x \in S^2$ and $l \in \mathbb{N}$ as in the lemma. Let $X^N \in \mathbf{X}^N$ be an N-tile containing $f^l(x)$. By Lemma 5.49, there exists an N-vertex $v^N \in \mathbf{V}^N \cap X^N$ such that

$$U^{N+\kappa}\left(f^l(x)\right) \subseteq W^N\left(v^N\right). \tag{5.92}$$

There exist $X^{N+l} \in \mathbf{X}^{N+l}$ and $v^{N+l} \in \mathbf{V}^{N+l} \cap X^{N+l}$ such that $x \in X^{N+l}$, $f^l\left(X^{N+l}\right) = X^N$, and $f^l\left(v^{N+l}\right) = v^N$. Since $l > M$ and $W^{N+l}\left(v^{N+l}\right) \subseteq U^{N+l}(x)$, we get from (5.90), (5.91), and (5.87) that if $z \in W^{N+l}\left(v^{N+l}\right)$, then

$$\begin{aligned} d\left(f^l(x), z\right) &\leq d\left(f^l(x), x\right) + d(x, z) \leq \delta + 2K\Lambda^{-(N+l)} \\ &\leq \frac{\Lambda^{-(N+\kappa)}}{2K} + \frac{2K\Lambda^{-(N+\kappa)}}{10K^2} \leq K^{-1}\Lambda^{-(N+\kappa)}. \end{aligned}$$

Thus by (5.91) and (5.92), we get

$$\overline{W}^{N+l}\left(v^{N+l}\right) \subseteq U^{N+\kappa}\left(f^l(x)\right) \subseteq W^N\left(v^N\right).$$

By Lemma 5.48, there exists $y \in \overline{W}^{N+l}\left(v^{N+l}\right) \subseteq U^{N+l}(x)$ such that $f^l(y) = y$.

It suffices now to verify that $d\left(f^i(x), f^i(y)\right) \leq \beta_0\delta\Lambda^{-(l-i)}$ for $i \in \{0, 1, \ldots, l\}$. Indeed, since by Proposition 2.6,

$$\{f^i(x), f^i(y)\} \subseteq \overline{W}^{N+l-i}\left(f^i\left(v^{N+l}\right)\right) \subseteq U^{N+l-i}\left(f^i\left(v^{N+l}\right)\right)$$

for $i \in \{0, 1, \ldots, l\}$, we get from (5.91), (5.89), and (5.88) that

$$d\left(f^i(x), f^i(y)\right) \leq 2K\Lambda^{-(N+l-i)} \leq \beta\Lambda^{-(l-i)} = \beta_0\delta\Lambda^{-(l-i)}.$$

□

The next lemma follows from the *topological transitivity* (see [PU10, Definition 4.3.1]) of expanding Thurston maps and Lemma 4.3.4 in [PU10]. We include a direct proof here for completeness.

Lemma 5.51 *Let $f: S^2 \to S^2$ be an expanding Thurston map. Then there exists a point $x \in S^2$ such that the set $\{f^n(x) \mid n \in \mathbb{N}\}$ is dense in S^2.*

Proof By Theorem 9.1 in [BM17], the topological dynamical system (S^2, f) is a factor of the topological dynamical system (J^ω, Σ) of the left-shift Σ on the space J^ω of all infinite sequences in a finite set J of cardinality card $J = \deg f$. More precisely, if we equip $J^\omega = \prod_{i=1}^{+\infty} J$ with the product topology, where $J = \{1, 2, \ldots, \deg f\}$, and let the left-shift operator Σ map $(i_1, i_2, \dots) \in J^\omega$ to $(i_2, i_3, \dots)$, then there exists a surjective continuous map $\xi\colon J^\omega \to S^2$ such that $\xi \circ \Sigma = f \circ \xi$.

It suffices now to find $y \in J^\omega$ such that the set $\{\Sigma^n(y) \mid n \in \mathbb{N}\}$ is dense in J^ω. Indeed, if we let $\{w_i\}_{i\in\mathbb{N}}$ be an enumeration of all elements in the set $\bigcup_{i=1}^{+\infty} J^i$ of all finite sequences in J, and set y to be the concatenation of $w_1, w_2, \dots$, then it is clear that $\{\Sigma^n(y) \mid n \in \mathbb{N}\}$ is dense in J^ω. □

Following similar argument as in the proof of Proposition 4.4.5 in [PU10], we get the next proposition. Note that here we do not explicitly use the distortion lemmas (Lemma 5.3 and Lemma 5.4).

Proposition 5.52 *Let f, $\mathscr{C}$, d, Λ satisfy the Assumptions. Let $\phi, \psi \in C^{0,\alpha}((S^2, d), \mathbb{C})$ be complex-valued Hölder continuous functions with an exponent $\alpha \in (0, 1]$. If $f(\mathscr{C}) \subseteq \mathscr{C}$, then the following conditions are equivalent:*

(i) *If $x \in S^2$ satisfies $f^n(x) = x$ for some $n \in \mathbb{N}$, then $S_n\phi(x) = S_n\psi(x)$.*
(ii) *There exists a constant $C > 0$ such that $|S_n\phi(x) - S_n\psi(x)| \leq C$ for $x \in S^2$ and $n \in \mathbb{N}_0$.*
(iii) *There exists $u \in C^{0,\alpha}((S^2, d), \mathbb{C})$ such that $\phi - \psi = u \circ f - u$.*

Moreover, if $\phi, \psi \in C^{0,\alpha}(S^2, d)$ are real-valued, then we can also choose u in (iii) *to be real-valued.*

Proof The implication from (iii) to (ii) holds since $|S_n\phi(x) - S_n\psi(x)| = |(u \circ f^n)(x) - u(x)| \leq 2\,\|u\|_\infty$ for $x \in S^2$ and $n \in \mathbb{N}$.

To prove that (ii) implies (i), we suppose that $f^n(x) = x$ and $D = S_n\phi(x) - S_n\psi(x) \neq 0$ for some $x \in S^2$ and some $n \in \mathbb{N}$. Then $|S_{ni}\phi(x) - S_{ni}\psi(x)| = iD > C$ for i large enough, contradicting (ii).

We now prove the implication from (i) to (iii).

Let $x \in S^2$ be a point from Lemma 5.51 so that the set $A = \{f^i(x) \mid i \in \mathbb{N}\}$ is dense in S^2. Set $x_i = f^i(x)$ for $i \in \mathbb{N}$. Note that $x_i \neq x_j$ for $j > i \geq 0$. Denote $\eta = \phi - \psi$. Then $\eta \in C^{0,\alpha}(S^2, d)$. We define a function v on A by setting $v(x_n) = S_n\eta(x)$. We will prove that v extends to a Hölder continuous function $u \in C^{0,\alpha}((S^2, d), \mathbb{C})$ defined on S^2 by showing that v is Hölder continuous with an exponent α on A.

Fix some $n, m \in \mathbb{N}$ with $n < m$ and $d(x_n, x_m) < \frac{1}{2}\delta_0$, where $\delta_0 \in (0, 1)$ is a constant depending only on f, $\mathscr{C}$, and d from Lemma 5.50. Set $\varepsilon = d(x_n, x_m)$. We can choose $k \in \mathbb{N}$ such that $d(x_m, x_k) < \varepsilon$ and $k > m + M$, where $M \in \mathbb{N}_0$ is a constant from Lemma 5.50. Note that $d(x_n, x_k) \leq d(x_n, x_m) + d(x_m, x_k) < 2\varepsilon < \delta_0$ and $k > n + M$. Thus by applying Lemma 5.50 with $\delta = 2\varepsilon$, there exist periodic points $p, q \in S^2$ such that $f^{k-n}(p) = p$, $f^{k-m}(q) = q$, $d\left(f^i(x_n), f^i(p)\right) < \beta_0\delta\Lambda^{-(k-n-i)}$ for $i \in \{0, 1, \dots, k-n\}$, and $d\left(f^j(x_m), f^j(q)\right) < \beta_0\delta\Lambda^{-(k-m-j)}$ for $j \in \{0, 1, \dots, k-m\}$, where $\beta_0 > 0$ is a constant depending only on f, $\mathscr{C}$, and d from Lemma 5.49. Then by (i), we get that

$$
\begin{aligned}
|v(x_n) - v(x_m)| &= |S_n\eta(x) - S_m\eta(x)| \le |S_{k-n}\eta(x_n)| + |S_{k-m}\eta(x_m)| \\
&= |S_{k-n}\eta(x_n) - S_{k-n}\eta(p)| + |S_{k-m}\eta(x_m) - S_{k-m}\eta(q)| \\
&\le \sum_{i=0}^{k-n-1} \left|\eta\left(f^i(x_n)\right) - \eta\left(f^i(p)\right)\right| + \sum_{j=0}^{k-m-1} \left|\eta\left(f^j(x_m)\right) - \eta\left(f^j(q)\right)\right| \\
&\le |\eta|_\alpha \beta_0^\alpha \delta^\alpha \left(\sum_{i=0}^{k-n-1} \Lambda^{-\alpha(k-n-i)} + \sum_{j=0}^{k-m-1} \Lambda^{-\alpha(k-m-i)} \right) \\
&\le 2^{1+\alpha} |\eta|_\alpha \beta_0^\alpha \varepsilon^\alpha \sum_{i=0}^{\infty} \Lambda^{-\alpha i} = C d(x_n, x_m)^\alpha,
\end{aligned}
$$

where $C = 2^{1+\alpha}(1 - \Lambda^{-\alpha})^{-1} |\eta|_\alpha \beta_0^\alpha$ is a constant depending only on f, $\mathscr{C}$, d, η, and α. It immediately follows that v extends continuously to a Hölder continuous function $u \in C^{0,\alpha}((S^2, d), \mathbb{C})$ with an exponent α defined on $\overline{A} = S^2$. Note that if ϕ and ψ are real-valued, so is u. Since $u|_A = v$ and

$$(v \circ f)(x_i) - v(x_i) = v(x_{i+1}) - v(x_i) = S_{i+1}\eta(x) - S_i\eta(x) = \eta(f^i(x)) = \phi(x_i) - \psi(x_i),$$

for $i \in \mathbb{N}$, we get that $(u \circ f)(y) - u(y) = \phi(y) - \psi(y)$ for $y \in S^2$ by continuity. □

Lemma 5.53 *Let f, d satisfy the Assumptions. Let $\phi, \psi, u \in C((S^2, d), \mathbb{C})$ be continuous functions. Then the following are equivalent:*

(i) *$\phi - \psi = u \circ f - u + K_1$ for some constant $K_1 \in \mathbb{C}$.*

(ii) *For all $n \in \mathbb{N}$, there exists a constant $K_2 \in \mathbb{C}$ such that $S_n\phi - S_n\psi = u \circ f^n - u + K_2$.*

(iii) *There exist constants $n \in \mathbb{N}$ and $K_3 \in \mathbb{C}$ such that $S_n\phi - S_n\psi = u \circ f^n - u + K_3$.*

Proof It is clear that (i) implies (ii) and that (ii) implies (iii). To verify that (iii) implies (i), we assume that

$$S_n\phi(x) - S_n\psi(x) = u \circ f^n(x) - u(x) + K_3 \tag{5.93}$$

for $x \in S^2$.

Fix an arbitrary point $y \in S^2$. By subtracting (5.93) with $x = y$ from (5.93) with $x = f(y)$, we get

$$
\begin{aligned}
&(u \circ f^{n+1})(y) - (u \circ f^n)(y) + (u \circ f)(y) - u(y) \\
&= (\phi \circ f^n)(y) - \phi(y) - (\psi \circ f^n)(y) + \psi(y),
\end{aligned}
$$

or equivalently,

$$\phi(f^n(y)) - \psi(f^n(y)) - (u \circ f)(f^n(y)) + u(f^n(y)) = \phi(y) - \psi(y) - (u \circ f)(y) + u(y). \tag{5.94}$$

Let $z \in S^2$ be a point from Lemma 5.51 so that the set $A = \{f^{ni}(z) \mid i \in \mathbb{N}\}$ is dense in S^2. By replacing y in (5.94) with $f^{ni}(z)$ for $i \in \mathbb{N}_0$ and induction, we get that

$$\phi(f^{ni}(z)) - \psi(f^{ni}(z)) - (u \circ f)(f^{ni}(z)) + u(f^{ni}(z)) = K_1$$

for $i \in \mathbb{N}$, where $K_1 = \phi(z) - \psi(z) - (u \circ f)(z) + u(z)$. Since A is dense in S^2, we get that $\phi(x) - \psi(x) - (u \circ f)(x) + u(x) = K_1$ for $x \in S^2$. □

We are now ready to prove Theorem 5.45.

Proof (*Proof of Theorem* 5.45) We fix a Jordan curve $\mathscr{C} \subseteq S^2$ that satisfies the Assumptions (see Theorem 2.16 for the existence of such $\mathscr{C}$).

We first prove the backward implication. We assume that

$$\phi - \psi - K\mathbb{1}_{S^2} = u \circ f - u \tag{5.95}$$

for some $u \in C(S^2)$ and $K \in \mathbb{R}$. It follows immediately from Proposition 5.19 that

$$P(f, \phi) = P(f, \psi) + K. \tag{5.96}$$

By Theorem 5.17, Proposition 5.19, Corollary 5.20, and Theorem 5.36, the measure μ_ϕ (resp. μ_ψ) is a Gibbs state with respect to f, $\mathscr{C}$, and ϕ (resp. ψ) with constants $P_{\mu_\phi} = P(f, \phi)$ and C_{μ_ϕ} (resp. $P_{\mu_\psi} = P(f, \psi)$ and C_{μ_ψ}). Then by (5.8), (5.95), and (5.96), for $i \in \mathbb{N}_0$ and $X^i \in \mathbf{X}^i(f, \mathscr{C})$,

$$\begin{aligned} \frac{\mu_\phi(X^i)}{\mu_\psi(X^i)} &\leq C_{\mu_\phi} C_{\mu_\psi} \frac{\exp(S_i\psi(x) - iP(f, \psi))}{\exp(S_i\phi(x) - iP(f, \phi))} \\ &= C_{\mu_\phi} C_{\mu_\psi} \exp(u(x) - (u \circ f)(x)) \leq C_{\mu_\phi} C_{\mu_\psi} \exp(2\,\|u\|_\infty), \end{aligned} \tag{5.97}$$

where $x \in X^i$. Let $E \subseteq S^2$ be a Borel set with $\mu_\psi(E) = 0$. Fix an arbitrary number $\varepsilon > 0$. We can find an open set $U \subseteq S^2$ such that $E \subseteq U$ and $\mu_\psi(U) < \varepsilon$. Set

$$V = \bigcup \left\{ \operatorname{inte}(X) \,\middle|\, X \in \bigcup_{i=0}^{+\infty} \mathbf{V}^i(f, \mathscr{C}),\ X \cap E \neq \emptyset,\ X \subseteq U \right\}.$$

Then $E \subseteq V \cup A$, where $A = \bigcup\limits_{i=0}^{+\infty} f^{-i}(\mathscr{C})$. By Proposition 5.39, we have $\mu_\phi(A) = \mu_\psi(A) = 0$. So by (5.97), we get

$$\mu_\phi(E) \leq \mu_\phi(V) \leq D\mu_\psi(V) \leq D\mu_\psi(U) < D\varepsilon,$$

where $D = C_{\mu_\phi} C_{\mu_\psi} \exp(2 \|u\|_\infty)$. Thus μ_ϕ is absolutely continuous with respect to μ_ψ. Similarly μ_ψ is absolutely continuous with respect to μ_ϕ. On the other hand, by Corollary 5.44, both μ_ϕ and μ_ψ are ergodic measures. So suppose $\mu_\phi \neq \mu_\psi$, then they must be mutually singular (see for example, [Wal82, Theorem 6.10(iv)]). Hence $\mu_\phi = \mu_\psi$.

We will now prove the forward implication. We assume $\mu_\phi = \mu_\psi$.

Denote $F = f^n$, where $n = n_{\mathscr{C}}$ is a number from the Assumptions with $f^n(\mathscr{C}) = F(\mathscr{C}) \subseteq \mathscr{C}$. By Remark 2.11 the map F is also an expanding Thurston map.

For the rest of the proof, we denote $S_m \eta = \sum_{i=0}^{m-1} \eta \circ f^i$ and $\widetilde{S}_m \eta = \sum_{i=0}^{m-1} \eta \circ F^i$ for $\eta \in C(S^2)$ and $m \in \mathbb{N}_0$.

Denote $\phi_n = S_n \phi$ and $\psi_n = S_n \psi$. It follows immediately from Lemma 2.15 that $\phi_n, \psi_n \in C^{0,\alpha}(S^2, d)$.

Note that since μ_ϕ is an equilibrium state for f and ϕ, it follows that μ_ϕ is also an equilibrium state for F and ϕ_n. Indeed, by (3.4) and the fact that $h_{\mu_\phi}(f^n) = n h_{\mu_\phi}(f)$ (see for example, [Wal82, Theorem 4.13]), we have

$$P_{\mu_\phi}(F, \phi_n) = h_{\mu_\phi}(f^n) + \int S_n \phi \, \mathrm{d}\mu_\phi = n h_{\mu_\phi}(f) + n \int \phi \, \mathrm{d}\mu_\phi = n P(f, \phi) = P(F, \phi_n),$$

where the last equality follows immediately from Proposition 5.19. Similarly, the measure $\mu_\phi = \mu_\psi$ is an equilibrium state for F and ψ_n.

Thus by Theorem 5.17, Proposition 5.19, Corollary 5.20, and Theorem 5.36, the measure $\mu_\phi = \mu_\psi$ is both a Gibbs state with respect to F, $\mathscr{C}$, and ϕ_n, and with constants $P(F, \phi_n)$ and C, as well as a Gibbs state with respect to F, $\mathscr{C}$, and ψ_n, and with constants $P(F, \psi_n)$ and C', for some $C \geq 1$ and $C' \geq 1$. By (5.8), we have

$$\frac{1}{CC'} \leq \exp\left(\widetilde{S}_m \phi_n(x) - \widetilde{S}_m \psi_n(x) - m P(F, \phi_n) + m P(F, \psi_n)\right) \leq CC'$$

for $x \in S^2$ and $m \in \mathbb{N}_0$. So $\left|\widetilde{S}_m \overline{\phi}_n(x) - \widetilde{S}_m \overline{\psi}_n(x)\right| \leq \log(CC')$ for $x \in S^2$ and $m \in \mathbb{N}_0$, where $\overline{\phi}_n(x) = \phi_n(x) - P(F, \phi_n) \in C^{0,\alpha}(S^2, d)$ and $\overline{\psi}_n(x) = \psi_n(x) - P(F, \psi_n) \in C^{0,\alpha}(S^2, d)$. By Proposition 5.52, there exists $u \in C^{0,\alpha}(S^2, d)$ such that

$$(u \circ f^n)(x) - u(x) = \overline{\phi}_n(x) - \overline{\psi}_n(x) = S_n \phi(x) - S_n \psi(x) - \delta$$

for $x \in S^2$, where $\delta = P(F, \phi_n) - P(F, \psi_n)$. By Lemma 5.53, we get $\phi - \psi = u \circ f - u + K$ for some constant $K \in \mathbb{R}$. □

5.6 Equidistribution

In this section, we will discuss equidistribution results for preimages. Let f, d, ϕ, α satisfy the Assumptions and let μ_ϕ be the unique equilibrium state for f and ϕ throughout this section. We prove in Proposition 5.54 three versions of equidistribution of preimages under f^n as $n \longrightarrow +\infty$ with respect to μ_ϕ and m_ϕ as defined in Corollary 5.32, respectively. Proposition 5.54 partially generalizes Theorem 4.2, where we established the equidistribution of preimages with respect to the measure of maximal entropy.

Proposition 5.54 *Let f, d, ϕ, α satisfy the Assumptions. Let μ_ϕ be the unique equilibrium state for f and ϕ, and m_ϕ be as in Corollary 5.32 and $\widetilde{\phi}$ as defined in (5.48). For each sequence $\{x_n\}_{n\in\mathbb{N}}$ of points in S^2, we define the Borel probability measures*

$$\xi_n = \frac{1}{Z_n(\phi)} \sum_{y\in f^{-n}(x_n)} \deg_{f^n}(y) \exp\left(S_n\phi(y)\right) \delta_y, \tag{5.98}$$

$$\widehat{\xi}_n = \frac{1}{Z_n(\phi)} \sum_{y\in f^{-n}(x_n)} \deg_{f^n}(y) \exp\left(S_n\phi(y)\right) \frac{1}{n} \sum_{i=0}^{n-1} \delta_{f^i(y)}, \tag{5.99}$$

$$\widetilde{\xi}_n = \frac{1}{Z_n(\widetilde{\phi})} \sum_{y\in f^{-n}(x_n)} \deg_{f^n}(y) \exp\left(S_n\widetilde{\phi}(y)\right) \delta_y, \tag{5.100}$$

for each $n \in \mathbb{N}_0$, where $Z_n(\psi) = \sum_{y\in f^{-n}(x_n)} \deg_{f^n}(y) \exp\left(S_n\psi(y)\right)$, for $\psi \in C(S^2)$. Then

$$\xi_n \xrightarrow{w^*} m_\phi \text{ as } n \longrightarrow +\infty, \tag{5.101}$$

$$\widehat{\xi}_n \xrightarrow{w^*} \mu_\phi \text{ as } n \longrightarrow +\infty, \tag{5.102}$$

$$\widetilde{\xi}_n \xrightarrow{w^*} \mu_\phi \text{ as } n \longrightarrow +\infty. \tag{5.103}$$

We note that when $\phi \equiv 0$ and $x_n = x_{n+1}$ for each $n \in \mathbb{N}$, the versions (5.101) and (5.103) reduce to (4.2) of Theorem 4.2.

Proof We note that (5.102) follows directly from Lemma 5.33.

The proof of (5.101) is similar to that of Lemma 5.33. For the completeness, we include it here in detail.

For each sequence $\{x_n\}_{n\in\mathbb{N}}$ of points in S^2, and each $u \in C(S^2, d)$, by (3.8) and (5.28) we have

$$\langle \xi_n, u \rangle = \frac{\mathcal{L}_\phi^n(u)(x_n)}{\mathcal{L}_\phi^n(\mathbb{1})(x_n)} = \frac{\mathcal{L}_{\overline{\phi}}^n(u)(x_n)}{\mathcal{L}_{\overline{\phi}}^n(\mathbb{1})(x_n)}.$$

By Theorem 5.31,

$$\left\| \mathcal{L}_{\overline{\phi}}^n(\mathbb{1}) - u_\phi \right\|_\infty \longrightarrow 0 \text{ and } \left\| \mathcal{L}_{\overline{\phi}}^n(u) - u_\phi \int u \,\mathrm{d}m_\phi \right\|_\infty \longrightarrow 0$$

as $n \longrightarrow +\infty$. So by (5.35),

$$\lim_{n\to+\infty} \frac{\mathscr{L}^n_{\widetilde{\phi}}(u)(x_n)}{\mathscr{L}^n_{\widetilde{\phi}}(\mathbb{1})(x_n)} = \int u \, \mathrm{d}m_\phi.$$

Hence, (5.101) holds.

Finally, (5.103) follows from (5.101) and the fact that $\widetilde{\phi} \in C^{0,\alpha}(S^2, d)$ (Lemma 5.26) and $m_{\widetilde{\phi}} = \mu_\phi$ (Corollary 5.32). □

5.7 A Random Iteration Algorithm

In this section, we follow the idea of [HT03] to prove that for each $p \in S^2$, the equilibrium state μ_ϕ for an expanding Thurston map f and a given real-valued Hölder continuous potential (with respect to a visual metric) is almost surely the limit of

$$\frac{1}{n}\sum_{i=0}^{n-1} \delta_{q_i}$$

as $n \longrightarrow +\infty$ in the weak* topology, where $q_0 = p$, and q_i is one of the points x in $f^{-1}(q_{i-1})$, chosen with probability $\deg_f(x) \exp\big(\widetilde{\phi}(x)\big)$, for each $i \in \mathbb{N}$. Here $\widetilde{\phi}$ is defined in (5.48). Note that when $\phi \equiv 0$, we have that μ_ϕ is the measure of maximal entropy of f and that $\widetilde{\phi} = -h_{\mathrm{top}}(f) = -\log(\deg f)$, thus $\deg_f(x) \exp\big(\widetilde{\phi}(x)\big) = \frac{\deg_f(x)}{\deg f}$.

To give a more precise formulation, we will use the language of Markov process from the probability theory (see for example, [Du10] for an introduction).

Let (X, d) be a compact metric space. Equip the space $\mathscr{P}(X)$ of Borel probability measures with the weak* topology. A continuous map $X \to \mathscr{P}(X)$ assigning to each $x \in X$ a measure μ_x defines a *random walk* on X. We define the corresponding *Markov operator* $Q\colon C(X) \to C(X)$ by

$$Q\phi(x) = \int \phi(y) \, \mathrm{d}\mu_x(y). \tag{5.104}$$

Let Q^* be the adjoint operator of Q, i.e., for each $\phi \in C(X)$ and $\rho \in \mathscr{P}(X)$,

$$\int Q\phi \, \mathrm{d}\rho = \int \phi \, \mathrm{d}(Q^*\rho). \tag{5.105}$$

Consider a stochastic process $(\Omega, \mathscr{F}, P)$, where

1. $\Omega = \{(\omega_0, \omega_1, \dots) \,|\, \omega_i \in X, i \in \mathbb{N}_0\} = \prod_{i=0}^{+\infty} X$, equipped with the product topology,

2. $\mathscr{F}$ is the Borel σ-algebra on Ω,
3. $P \in \mathscr{P}(\Omega)$.

This process is a *Markov process with transition probabilities* $\{\mu_x\}_{x \in X}$ if

$$P\{\omega_{n+1} \in A \mid \omega_0 = z_0, \omega_1 = z_1, \dots, \omega_n = z_n\} = \mu_{z_n}(A) \tag{5.106}$$

for all $n \in \mathbb{N}_0$, Borel subsets $A \subseteq X$, and $z_0, z_1, \dots, z_n \in X$.

The transition probabilities $\{\mu_x\}_{x \in X}$ are determined by the operator Q and so we can speak of a *Markov process determined by* Q.

Let f, d, ϕ, α satisfy the Assumptions. Set $Q = \mathscr{L}_{\widetilde{\phi}}$. Then for each $u \in C(S^2)$,

$$Qu(x) = \int u(y)\, \mathrm{d}\mu_x(y),$$

where

$$\mu_x = \sum_{z \in f^{-1}(x)} \deg_f(z) \exp\big(\widetilde{\phi}(z)\big)\delta_z.$$

By (5.51), we get that $\mu_x \in \mathscr{P}(S^2)$ for each $x \in S^2$. We showed that the Ruelle operator in (3.7) is well-defined, from which it immediately follows that the map $x \mapsto \mu_x$ from S^2 to $\mathscr{P}(S^2)$ is continuous with respect to weak* topology on $\mathscr{P}(S^2)$.

Fix an arbitrary $z \in S^2$. Then there exists a unique Markov process $(\Omega, \mathscr{F}, P_z)$ determined by Q with

1. $\Omega = \prod\limits_{i=0}^{+\infty} S^2$, equipped with the product topology,
2. $\mathscr{F}$ being the Borel σ-algebra on Ω,
3. P_z being a Borel probability measure on Ω satisfying

$$P_z\{\omega_{n+1} \in A \mid \omega_0 = z, \omega_1 = z_1, \dots, \omega_n = z_n\} = \mu_{z_n}(A)$$

for all $n \in \mathbb{N}$, Borel subset $A \subseteq S^2$, and $z_1, z_2, \dots, z_n \in S^2$.

The existence and uniqueness of P_z follows from [Lo77, Theorem 1.4.2]. Since the Markov process $(\Omega, \mathscr{F}, P_z)$ is determined by f and ϕ as well, we will also call $(\Omega, \mathscr{F}, P_z)$ the *Markov process determined by f and ϕ*.

Now we can formulate our main theorem for this section.

Theorem 5.55 *Let $f : S^2 \to S^2$ be an expanding Thurston map and d be a visual metric on S^2 for f. Let $\phi \in C^{0,\alpha}(S^2, d)$ be a real-valued Hölder continuous function with an exponent $\alpha \in (0, 1]$. Let μ_ϕ be the unique equilibrium state for f and ϕ. Let $(\Omega, \mathscr{F}, P_z)$ be the Markov process determined by f and ϕ. Then for each $z \in S^2$, we have that P_z-almost surely,*

$$\frac{1}{n}\sum_{j=0}^{n-1}\delta_{\omega_j} \xrightarrow{w^*} \mu_\phi \text{ as } n \longrightarrow +\infty. \tag{5.107}$$

In other words, if we fix a point $z \in S^2$ and set it as the first point in an infinite sequence, and choose each of the following points randomly according to the Markov process determined by f and ϕ, then P_z-almost surely, the probability measure equally distributed on the first n points in the sequence converges in the weak* topology to μ_f as $n \longrightarrow +\infty$.

In order to prove Theorem 5.55, we need a theorem of H. Furstenberg and Y. Kifer from [FK83].

Theorem 5.56 (H. Furstenberg and Y. Kifer) *Let $\Omega = \{\omega_n \in X \mid n \in \mathbb{N}_0\}$ be the Markov process determined by the operator Q. Assume that there exists a unique Borel probability measure μ that is invariant under the adjoint operator Q^* on $\mathcal{P}(X)$. Then for each $\omega_0 \in X$, we have that P_{ω_0}-almost surely,*

$$\frac{1}{n}\sum_{j=0}^{n-1}\delta_{\omega_j} \xrightarrow{w^*} \mu \text{ as } n \longrightarrow +\infty. \tag{5.108}$$

Theorem 5.55 follows immediately from Theorem 5.56 and the fact that the equilibrium state μ_ϕ is the unique Borel probability measure on S^2 that satisfies $\mathcal{L}^*_{\widetilde{\phi}}(\mu_\phi) = \mu_\phi$ (see Corollary 5.32).

Chapter 6
Asymptotic h-Expansiveness

This Chapter is devoted to the investigation of the weak expansion properties of expanding Thurston maps. The main theorem for this chapter is the following.

Theorem 6.1 *Let $f \colon S^2 \to S^2$ be an expanding Thurston map. Then f is asymptotically h-expansive if and only if f has no periodic critical points. Moreover, f is not h-expansive.*

As an immediate consequence of this theorem and J. Buzzi's result on the asymptotic h-expansiveness of C^∞-maps on compact Riemannian manifolds [Buz97], we get the following corollary, which partially answers a question of K. Pilgrim (see Problem 2 in [BM10, Sect. 21]).

Corollary 6.2 *An expanding Thurston map with at least one periodic critical point cannot be conjugate to a C^∞-map from the Euclidean 2-sphere to itself.*

Remark 6.3 Corollary 6.2 can also be proved using an elementary argument which we include here. Suppose that a C^∞-map $f : \widehat{\mathbb{C}} \to \widehat{\mathbb{C}}$ is an expanding Thurston map with a periodic critical point p. We can assume that $p = 0$. By Theorem 2.16, there exists $N \in \mathbb{N}$ such that the C^∞-map $F = f^N$ is an expanding Thurston map with a fixed critical point 0 such that there exists a Jordan curve $\mathscr{C} \subseteq S^2$ with $F(\mathscr{C}) \subseteq \mathscr{C}$ and $\operatorname{post} F \subseteq \mathscr{C}$. Then there exists $r > 0$ such that the Jacobian determinant $\det DF$ satisfies $|\det DF(z)| < \frac{1}{2}$ for all $z \in B_\rho(0, r)$. Here ρ is the Euclidean metric on $\mathbb{C}$. Since F is an expanding Thurston map, there exists $n \in \mathbb{N}$ such that the n-flower $W^n(0) \in \mathbf{W}^n(F, \mathscr{C})$ is a subset of $B_\rho(0, r)$. Note that $F(W^n(0)) = W^{n-1}(0)$. Thus $m(W^{n-1}(0)) \leq \int_{W^n(0)} |\det DF| \, \mathrm{d}m \leq \frac{1}{2} m(W^n(0))$, where m is the Lebesgue measure on $\mathbb{C}$. On the other hand, it is clear that $W^n(0) \subsetneq W^{n-1}(0)$ since F is an expanding Thurston map. Since flowers are open, we get that $m(W^n(0)) < m(W^{m-1}(0))$, a contradiction.

The next corollary follows from Theorem 6.1 and M. Misiurewicz's result in [Mis76] (see also the discussion in Chap. 1).

Z. Li, *Ergodic Theory of Expanding Thurston Maps*, Atlantis Studies in Dynamical Systems 4, DOI 10.2991/978-94-6239-174-1_6

Corollary 6.4 *Let $f : S^2 \to S^2$ be an expanding Thurston map without periodic critical points. Then the measure-theoretic entropy $h_\mu(f)$ considered as a function of μ on the space $\mathcal{M}(S^2, f)$ of f-invariant Borel probability measures is upper semi-continuous. Here $\mathcal{M}(S^2, f)$ is equipped with the weak* topology.*

Recall that if X is a metric space, a function $h : X \to [-\infty, +\infty]$ is *upper semi-continuous* if $\limsup_{y \to x} h(y) \leq h(x)$ for all $x \in \mathbf{X}$.

Note that Corollary 6.4 implies a partially stronger existence result than the one obtained in Theorem 5.1.

Theorem 6.5 *Let $f : S^2 \to S^2$ be an expanding Thurston map without periodic critical points and $\psi \in C(S^2)$ be a real-valued continuous function on S^2 (equipped with the standard topology). Then there exists at least one equilibrium state for the map f and the potential ψ.*

See the end of Sect. 6.3 for a quick proof of Theorem 6.5.

In Sect. 6.1, we prove three lemmas that will be used in the proof of the asymptotic h-expansiveness of expanding Thurston maps without periodic critical points. Lemma 6.6 states that any expanding Thurston map is uniformly locally injective away from the critical points, in the sense that if one fixes such a map f and a visual metric d on S^2 for f, then for each $\delta > 0$ sufficiently small and each $x \in S^2$, the map f is injective on the δ-ball centered at x as long as x is not in a $\tau(\delta)$-ball of any critical point of f, where $\tau(\delta)$ can be made arbitrarily small if δ tends to 0. In Lemma 6.7 we prove a few properties of flowers in the cell decompositions of S^2 induced by an expanding Thurston map and some special f-invariant Jordan curve. Lemma 6.8 gives a covering lemma to cover sets of the form $\bigcap\limits_{i=0}^{n} f^{-i}(W_i)$ by $(m+n)$-flowers, where $m \in \mathbb{N}_0$, $n \in \mathbb{N}$, and each W_i is an m-flower.

We review some basic concepts from graph theory in Sect. 6.2, and provide simple upper bounds of the numbers of leaves of certain trees in Lemma 6.9. Note that we will not use any nontrivial facts from graph theory in this monograph.

Section 6.3 consists of the proof of Theorem 6.1 in the form of three separate theorems. Namely, we show in Theorem 6.10 the asymptotic h-expansiveness of expanding Thurston maps without periodic critical points. The proof relies on a quantitative upper bound of the frequency for an orbit under such a map to get close to the set of critical points. Lemma 6.9 and terminology from graph theory is used here to make the statements in the proof precise. We then prove in Theorems 6.11 and 6.13 the lack of asymptotic h-expansiveness of expanding Thurston maps with periodic critical points and the lack of h-expansiveness of expanding Thurston maps without periodic critical points, respectively, by explicit constructions of periodic sequences $\{v_i\}_{i \in \mathbb{N}}$ of m-vertices for which one can give lower bounds for the numbers of open sets in the open cover $\bigvee\limits_{j=0}^{n-1} f^{-j}(\mathbf{W}^m)$ needed to cover the set $\bigcap\limits_{j=0}^{n-1} f^{-j}(W^m(v_{n-j}))$, for $l, m, n \in \mathbb{N}$ sufficiently large. Here $W^m(v_{n-j})$ denotes the m-flower of v_{n-j} (see (2.5)), and $\mathbf{W}^m$ is the set of all m-flowers (see (2.6)). These lower bounds lead to the conclusion that the topological tail entropy and topological conditional entropy, respectively, are strictly positive, proving the corresponding theorems (compare with Definitions 3.5 and 3.6). The periodic sequence $\{v_i\}_{i\in\mathbb{N}}$ of m-vertices in the proof of Theorem 6.11 shadows a certain infinite backward pseudo-orbit in such a way that each period of $\{v_i\}_{i\in\mathbb{N}}$ begins with a backward orbit starting at a critical point p which is a fixed point of f, and approaching p as the index i increases, and then ends with a constant sequence staying at p. The fact that the constant part of each period of $\{v_i\}_{i\in\mathbb{N}}$ can be made arbitrarily long is essential here and is not true if f has no periodic critical points. The periodic sequence $\{v_i\}_{i\in\mathbb{N}_0}$ of m-vertices in the proof of Theorem 6.13 shadows a certain infinite backward pseudo-orbit in such a way that each period of $\{v_i\}_{i\in\mathbb{N}_0}$ begins with a backward orbit starting at $f(p)$ and p, and approaching $f(p)$ as the index i increases, and then ends with $f(p)$. In this case p is a critical point whose image $f(p)$ is a fixed point. In both constructions, we may need to consider an iterate of f for the existence of p with the required properties. Combining Theorems 6.10, 6.11, and 6.13, we get Theorem 6.1. This chapter ends with a quick proof of Theorem 6.5, which asserts the existence of equilibrium states for expanding Thurston maps without periodic critical points and given continuous potentials.

6.1 Some Properties of Expanding Thurstons Maps

We need the following three lemmas for the proof of the asymptotic h-expansiveness of expanding Thurston maps with no periodic critical points.

Lemma 6.6 (Uniform local injectivity away from the critical points) *Let f, d satisfy the Assumptions. Then there exist a number $\delta_0 \in (0, 1]$ and a function $\tau\colon (0, \delta_0] \to (0, +\infty)$ with the following properties:*

(i) $\lim\limits_{\delta\to 0} \tau(\delta) = 0$.
(ii) For each $\delta \leq \delta_0$, the map f restricted to any open ball of radius δ centered outside the $\tau(\delta)$-neighborhood of crit f *is injective, i.e., $f|_{B_d(x,\delta)}$ is injective for each $x \in S^2 \setminus N_d^{\tau(\delta)}(\text{crit } f)$.*

This lemma is straightforward to verify, but for the sake of completeness, we include the proof here.

Proof We first define a function $r\colon S^2 \setminus \text{crit } f \to (0, +\infty)$ in the following way

$$r(x) = \sup\{R > 0 \mid f|_{B_d(x,R)} \text{ is injective}\},$$

for $x \in S^2 \setminus \text{crit } f$. Note that $r(x) \leq d(x, \text{crit } f) < +\infty$ for each $x \in S^2 \setminus \text{crit } f$. We also observe that the supremum is attained, since otherwise, suppose $f(y) = f(z)$ for some $y, z \in B_d(x, r(x))$, then f is not injective on the ball $B_d(x, R_0)$ containing y and z with $R_0 = \frac{1}{2}(r(x) + \max\{d(x, y), d(x, z)\}) < r(x)$, a contradiction.

We claim that r is continuous.

Indeed, we observe that for each pair of distinct points $x, y \in S^2$, we have $r(x) \geq r(y) - d(x, y)$. This is true since if $r(y) - d(x, y) > 0$, then $B_d(x, r(y) - d(x, y)) \subseteq B_d(y, r(y))$. Now by symmetry, $r(y) \geq r(x) - d(x, y)$. So $|r(x) - r(y)| \leq d(x, y)$, and the claim follows.

Next, we fix a sufficiently small number $t_0 > 0$ with $S^2 \setminus N_d^{t_0}(\text{crit } f) \neq \emptyset$. We define a function $\sigma : (0, t_0] \to (0, +\infty)$ by setting

$$\sigma(t) = \inf\{r(x) \mid x \in S^2 \setminus N_d^t(\text{crit } f)\}$$

for $t \in (0, t_0]$. We observe that σ is continuous and non-decreasing. Since $r(x) \leq d(x, \text{crit } f)$ for each $x \in S^2 \setminus \text{crit } f$, we can conclude that $\lim\limits_{t \to 0} \sigma(t) = 0$. By the definition of σ, we get that $f|_{B_d(x, \sigma(t))}$ is injective, for $t \in (0, t_0]$ and $x \in S^2 \setminus N_d^t(\text{crit } f)$.

Finally, we construct $\tau : (0, \delta_0] \to (0, +\infty)$, where $\delta_0 = \min\{1, \sigma(t_0)\}$ by setting

$$\tau(\delta) = \inf\{t \in (0, t_0] \mid \sigma(t) \geq \delta\} \tag{6.1}$$

for each $\delta \in (0, \delta_0]$. We note that $\lim\limits_{\delta \to 0} \tau(\delta) = 0$.

For $\delta \in (0, \delta_0]$ and $t \in (\tau(\delta), t_0]$, we have $\sigma(t) \geq \delta$ by (6.1) and the fact that σ is non-decreasing. Since σ is continuous on $(0, t_0]$, we get $\sigma(\tau(\delta)) \geq \delta$. For each $x \in S^2 \setminus N_d^{\tau(\delta)}(\text{crit } f)$, we know from the definition of σ that $f|_{B_d(x, \sigma(\tau(\delta)))}$ is injective. Therefore $f|_{B_d(x,\delta)}$ is injective. □

Lemma 6.7 *Let f and $\mathscr{C}$ satisfy the Assumptions. Fix $m, n \in \mathbb{N}_0$ with $m < n$. If $f(\mathscr{C}) \subseteq \mathscr{C}$ and no 1-tile in $\mathbf{X}^1(f, \mathscr{C})$ joins opposite sides of $\mathscr{C}$, then the following statements hold:*

(i) For each n-vertex $v \in \mathbf{V}^n(f, \mathscr{C})$ and each m-vertex $w \in \mathbf{V}^m(f, \mathscr{C})$, if $v \notin \overline{W}^m(w)$, then $W^m(w) \cap W^n(v) = \emptyset$.

(ii) For each n-tile $X^n \in \mathbf{X}^n(f, \mathscr{C})$, there exists an m-vertex $v^m \in \mathbf{V}^m(f, \mathscr{C})$ such that $X^n \subseteq W^m(v^m)$.

(iii) For each pair of distinct m-vertices $p, q \in \mathbf{V}^m(f, \mathscr{C})$, $\overline{W}^{n+1}(p) \cap \overline{W}^{n+1}(q) = \emptyset$.

Recall that W^n is defined in (2.5) and $\overline{W}^n(p)$ is the closure of $W^n(p)$. Note that a flower is an open set (see [BM17, Lemma 5.28]) and by definition a tile is a closed set.

Proof We first observe that in order to prove any of the statements in the lemma, it suffices to assume $n = m + 1$. So we will assume, without loss of generality, that $n = m + 1$.

(i) Since $v \notin \overline{W}^m(w)$, by (2.5) we get that $v \notin c$ for each m-cell $c \in \mathbf{D}^m$ with $w \in c$. Since $f(\mathscr{C}) = \mathscr{C}$, for each n-cell $c' \in \mathbf{D}^n$ and each m-cell $c \in \mathbf{D}^m$, if $c \cap \text{inte}(c') \neq \emptyset$, then $c' \subseteq c$ (see Lemma 4.3 and the proof of Lemma 5.7 in [BM17]). Thus $c \cap \text{inte}(c') = \emptyset$ for $c \in \mathbf{D}^m$ and $c' \in \mathbf{D}^n$ with $w \in c$ and $v \in c'$. So $W^m(w) \cap W^n(v) = \emptyset$ by (2.5).

(ii) Let $X^m \in \mathbf{X}^m$ be the unique m-tile with $X^n \subseteq X^m$. Depending on the location of X^n in X^m, it suffices to prove statement (ii) in the following cases:

(1) Assume that $X^n \subseteq \mathrm{inte}(X^m)$. Then $X^n \subseteq W^m(v^m)$ for any $v^m \in X^m \cap \mathbf{V}^m$.
(2) Assume that $\emptyset \neq X^n \cap e \subseteq \mathrm{inte}(e)$ for some m-edge $e \in \mathbf{E}^m$ with $e \subseteq X^m$. Then since no 1-tile joins opposite sides of $\mathscr{C}$, by Proposition 2.6(i), either $X^n \cap \partial X^m \subseteq \mathrm{inte}(e)$ or there exists $e' \in \mathbf{E}^m$ such that $X^n \cap \partial X^m \subseteq \mathrm{inte}(e) \cup \mathrm{inte}(e')$ and $e \cap e' = \{v\}$ for some $v \in \mathbf{V}^m$. In the former case, choose any $v^m \in e \cap \mathbf{V}^m$; and in the latter case, let $v^m = v$. Then $X^n \subseteq W^m(v^m)$.
(3) Assume $X^n \cap \mathbf{V}^m \neq \emptyset$. Since no 1-tile joins opposite sides of $\mathscr{C}$, by Proposition 2.6(i), there exists some m-vertex $v^m \in \mathbf{V}^m$ such that $X^n \cap \mathbf{V}^m = \{v^m\}$. Let $e, e' \in \mathbf{E}^m$ be the two m-edges that satisfy $e \cup e' \subseteq X^m$ and $e \cap e' = \{v^m\}$. Then by Proposition 2.6(i) and the assumption that no 1-tile joins opposite sides of $\mathscr{C}$, we get that $X^n \cap \partial X^m \subseteq \{v^m\} \cup \mathrm{inte}(e) \cup \mathrm{inte}(e')$. Thus $X^n \subseteq W^m(v^m)$.

(iii) We observe that since no 1-tile in $\mathbf{X}^1$ joins opposite sides of $\mathscr{C}$ and $f(\mathscr{C}) \subseteq \mathscr{C}$, by Proposition 2.6(i), each $(k+1)$-tile X^{k+1} contains at most one k-vertex, for $k \in \mathbb{N}_0$. Let $p, q \in \mathbf{V}^m$ be distinct. Then by Remark 2.7 and the observation above, we know $q \notin \overline{W}^n(p)$. So by part (i), we get $W^n(p) \cap W^{n+1}(q) = \emptyset$. Since flowers are open sets, we have $W^n(p) \cap \overline{W}^{n+1}(q) = \emptyset$. It suffices to prove that $\overline{W}^{n+1}(p) \subseteq W^n(p)$. Indeed this inclusion is true; for otherwise, there exist an $(n+1)$-tile $X^{n+1} \subseteq \overline{W}^{n+1}(p)$ and a point $x \in \overline{W}^n(p) \setminus W^n(p)$ such that $\{x, p\} \subseteq X^{n+1}$. By (2.5) and applying Proposition 2.6(i), we get a contradiction to the assumption that no 1-tile in $\mathbf{X}^1$ joins opposite sides of $\mathscr{C}$. $\square$

Let $f: S^2 \to S^2$ be an expanding Thurston map, and $\mathscr{C} \subseteq S^2$ a Jordan curve containing post f such that $f(\mathscr{C}) \subseteq \mathscr{C}$. We denote, for $m \in \mathbb{N}_0$, $n \in \mathbb{N}$, $q \in S^2$, and $q_i \in \mathbf{V}^m(f, \mathscr{C})$ for $i \in \{0, 1, \dots, n-1\}$,

$$\begin{aligned} E_m(q_0, q_1, \dots, q_{n-1}; q) &= \left\{ x \in f^{-n}(q) \,\middle|\, f^i(x) \in \overline{W}^m(q_i), i \in \{0, 1, \dots, n-1\} \right\} \\ &= f^{-n}(q) \cap \left(\bigcap_{i=0}^{n-1} f^{-i}\left(\overline{W}^m(q_i)\right) \right), \end{aligned} \tag{6.2}$$

where $\overline{W}^m(q_i)$ is the closure of the m-flower $W^m(q_i)$ as defined in Sect. 2.2.

Lemma 6.8 *Let $f: S^2 \to S^2$ be an expanding Thurston map, and $\mathscr{C} \subseteq S^2$ a Jordan curve containing* post f *such that $f(\mathscr{C}) \subseteq \mathscr{C}$. Then*

$$\bigcap_{i=0}^{n} f^{-i}(W^m(p_i)) \subseteq \bigcup_{x \in E_m(p_0, p_1, \dots, p_{n-1}; p_n)} W^{m+n}(x), \tag{6.3}$$

for $m \in \mathbb{N}_0$, $n \in \mathbb{N}$, and $p_i \in \mathbf{V}^m(f, \mathscr{C})$ for $i \in \{0, 1, \dots, n\}$. Here E_m is defined in (6.2).

Proof We prove the lemma by induction on $n \in \mathbb{N}$.

For $n = 1$, we know that for all $p_0, p_1 \in \mathbf{V}^m(f, \mathscr{C})$,

$$\begin{aligned} W^m(p_0) \cap f^{-1}(W^m(p_1)) &\subseteq \bigcup \{ W^{m+1}(x) \,|\, x \in f^{-1}(p_1), x \in \overline{W}^m(p_0) \} \\ &= \bigcup_{x \in E_m(p_0; p_1)} W^{m+1}(x) \end{aligned}$$

by (6.2) and the fact that $W^{m+1}(x) \cap W^m(p_0) = \emptyset$ if both $x \in \mathbf{V}^{m+1}(f, \mathscr{C})$ and $x \notin \overline{W}^m(p_0)$ are satisfied (see Lemma 6.7(i)).

We now assume that the lemma holds for $n = l$ for some $l \in \mathbb{N}$.

We fix a point $p_i \in \mathbf{V}^m(f, \mathscr{C})$ for each $i \in \{0, 1, \dots, l, l+1\}$. Then

$$\bigcap_{i=0}^{l+1} f^{-i}(W^m(p_i)) = W^m(p_0) \cap f^{-1}\left(\bigcap_{i=1}^{l+1} f^{-(i-1)}(W^m(p_i)) \right).$$

By induction hypothesis, the right-hand side of the above equation is a subset of

$$\begin{aligned} & W^m(p_0) \cap f^{-1}\left(\bigcup_{x \in E_m(p_1, p_2, \dots, p_l; p_{l+1})} W^{m+l}(x) \right) \\ = & \bigcup_{x \in E_m(p_1, p_2, \dots, p_l; p_{l+1})} \left(W^m(p_0) \cap f^{-1}\left(W^{m+l}(x) \right) \right) \\ \subseteq & \bigcup_{x \in E_m(p_1, p_2, \dots, p_l; p_{l+1})} \left(\bigcup \{ W^{m+l+1}(y) \,|\, y \in f^{-1}(x), y \in \overline{W}^m(p_0) \} \right) \\ = & \bigcup_{x \in E_m(p_1, p_2, \dots, p_l; p_{l+1})} \bigcup_{y \in E_m(p_0; x)} W^{m+l+1}(y), \end{aligned}$$

where the last two lines is due to (6.2) and the fact that $W^{m+l+1}(y) \cap W^m(p_0) = \emptyset$ if both $y \in \mathbf{V}^{m+l+1}(f, \mathscr{C})$ and $y \notin \overline{W}^m(p_0)$ are satisfied (see Lemma 6.7(i)).

We claim that

$$\bigcup_{x \in E_m(p_1, p_2, \dots, p_l; p_{l+1})} E_m(p_0; x) = E_m(p_0, p_1, \dots, p_l; p_{l+1}).$$

Assuming the claim, we then get

$$\bigcap_{i=0}^{l+1} f^{-i}(W^m(p_i)) \subseteq \bigcup_{x \in E_m(p_0, p_1, \dots, p_l; p_{l+1})} W^{m+l+1}(y).$$

Thus it suffices to prove the claim now. Indeed, by (6.2),

$$\bigcup_{x \in E_m(p_1, p_2, \ldots, p_l; p_{l+1})} E_m(p_0; x)$$

$$= \left\{ y \in f^{-1}(x) \,\middle|\, y \in \overline{W}^m(p_0), x \in f^{-l}(p_{l+1}) \cap \left(\bigcap_{i=1}^{l} f^{-i+1}(\overline{W}^m(p_i)) \right) \right\}$$

$$= \left\{ y \in f^{-l-1}(p_{l+1}) \,\middle|\, y \in \overline{W}^m(p_0), f(y) \in \bigcap_{i=1}^{l} f^{-i+1}(\overline{W}^m(p_i)) \right\}$$

$$= E_m(p_0, p_1, \ldots, p_l; p_{l+1}).$$

The induction step is now complete. □

6.2 Concepts from Graph Theory

We now review the notions of a simple directed graph and of a finite rooted tree that will be used in the proof of Theorem 6.10. Since the only purpose of such notions is to make the statements and proofs precise, and we will not use any nontrivial facts from graph theory, we adopt here a simplified approach to define relevant concepts as quickly as possible (compare [BJG09]).

A *simple directed graph* $\mathscr{G} = (\mathscr{V}(\mathscr{G}), \mathscr{E}(\mathscr{G}))$ is made up from a *set of vertices* $\mathscr{V}(\mathscr{G})$ and a *set of directed edges*

$$\mathscr{E}(\mathscr{G}) \subseteq \mathscr{V}(\mathscr{G}) \times \mathscr{V}(\mathscr{G}) \setminus \{(v, v) \mid v \in \mathscr{V}(\mathscr{G})\}.$$

A simple directed graph $\mathscr{G}$ is *finite* if card $\mathscr{V}(\mathscr{G}) < +\infty$. Two vertices $v, w \in \mathscr{V}(\mathscr{G})$ are *connected by a directed edge* (v, w) if $(v, w) \in \mathscr{E}(\mathscr{G})$. If $e = (v, w) \in \mathscr{E}(\mathscr{G})$, then we call v the *initial vertex* of e, denoted by $i(e)$, and w the *terminal vertex* of e, denoted by $t(e)$. The *indegree* of a vertex $v \in \mathscr{V}(\mathscr{G})$ is $\mathrm{d}^-(v) = \operatorname{card}\{w \in \mathscr{V}(\mathscr{G}) \mid (w, v) \in \mathscr{E}(\mathscr{G})\}$, and the *outdegree* of v is $d^+(v) = \operatorname{card}\{w \in \mathscr{V}(\mathscr{G}) \mid (v, w) \in \mathscr{E}(\mathscr{G})\}$. A *path* from a vertex $v \in \mathscr{V}(\mathscr{G})$ to a vertex $w \in \mathscr{V}(\mathscr{G})$ is a finite sequence of vertices $v = v_0, v_1, v_2, \ldots, v_{n-1}, v_n = w$ such that $(v_i, v_{i+1}) \in \mathscr{E}(\mathscr{G})$ for each $i \in \{0, 1, \ldots, n-1\}$. The *length* of such a path is n. The *distance from v to w* is the minimal length of all paths from v to w. By convention, the distance from v to v is 0, and if there is no path from v to w for $v \neq w$, then the distance from v to w is ∞. If the distance of v to w is $n \in \mathbb{N}_0$, then we say that w is at a distance n from v.

A finite simple directed graph $\mathscr{T}$ is a *finite rooted tree* if there exists a vertex $r \in \mathscr{V}(\mathscr{T})$ such that for each vertex $v \in \mathscr{V}(\mathscr{T}) \setminus \{r\}$ there exists a unique path from r to v. We call such a simple directed graph a *finite rooted tree with root* r, and r the *root* of $\mathscr{T}$. Note that a finite rooted tree has a unique root. A vertex v of a finite rooted tree $\mathscr{T}$ is called a *leaf* (of $\mathscr{T}$) if $d^+(v) = 0$. If $(v, w) \in \mathscr{E}(\mathscr{T})$, then w is said to be a *child* of v.

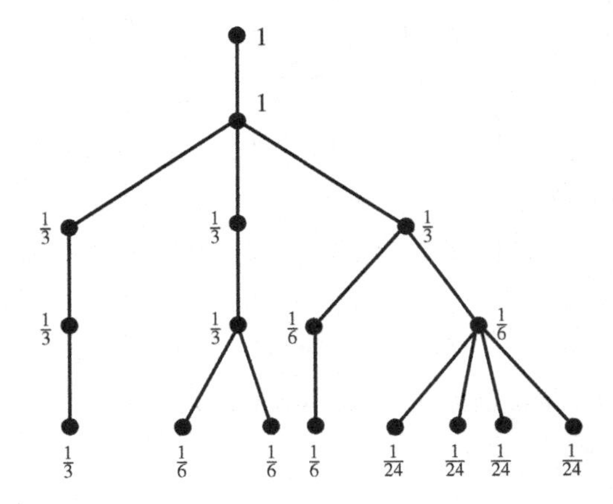

Fig. 6.1 The function h for a finite rooted tree

Lemma 6.9 (A bound for the number of leaves) *Let $\mathscr{T}$ be a finite rooted tree with root r whose leaves are all at the same distance from r. Assume that there exist constants $c, k \in \mathbb{N}$ with the following properties:*

(i) $d^+(x) \leq c$ for each vertex $x \in \mathscr{V}(\mathscr{T})$,
(ii) for each leaf v, the number of vertices w with $d^+(w) \geq 2$ in the path from r to v is at most k.

Then number of leaves of $\mathscr{T}$ is at most c^k.

Proof Let $N \in \mathbb{N}_0$ be the distance from r to any leaf of $\mathscr{T}$. For each $n \in \mathbb{N}_0$, we define $\mathscr{V}_n$ as the set of vertices of $\mathscr{T}$ at distance n from r. It is clear that a vertex $v \in \mathscr{V}(\mathscr{T})$ is a leaf of $\mathscr{T}$ if and only if $v \in \mathscr{V}_N$.

We can recursively construct a function $h \colon \mathscr{V}(\mathscr{T}) \to \mathscr{L}$ by setting $h(r) = 1$, and for each $v \in \mathscr{V}(\mathscr{T})$, defining $h(v) = \frac{h(w)}{d^+(w)}$, where $w \in \mathscr{V}(\mathscr{T})$ is the unique vertex with $(w, v) \in \mathscr{E}(\mathscr{T})$. See Fig. 6.1.

By the two properties in the hypothesis, we have $h(v) \geq c^{-k}$ for each leaf $v \in \mathscr{V}(\mathscr{T})$ of $\mathscr{T}$. On the other hand, it is easy to see from induction that $\sum\limits_{w \in \mathscr{V}_n} h(w) = 1$ for each $n \in \{0, 1, \dots, N\}$. In particular, we have $\sum\limits_{w \in \mathscr{V}_N} h(w) = 1$. Thus card $\mathscr{V}_N \leq c^k$. Therefore, the number of leaves of $\mathscr{T}$ is at most c^k. □

6.3 Proof of Theorem 6.1

We split Theorem 6.1 into three parts and prove each one separately here.

Theorem 6.10 *An expanding Thurston map $f \colon S^2 \to S^2$ with no periodic critical points is asymptotically h-expansive.*

Proof We need to show $h^*(f) = 0$. By (3.16), it suffices to prove that f^i is asymptotically h-expansive for some $i \in \mathbb{N}$. Note that by (2.2), f^i has no periodic critical points for each $i \in \mathbb{N}$ if f does not. Thus by Lemma 2.17, we can assume, without loss of generality, that there exists a Jordan curve $\mathscr{C} \subseteq S^2$ containing post f such that $f(\mathscr{C}) \subseteq \mathscr{C}$, and no 1-tile joins opposite sides of $\mathscr{C}$. We consider the cell decompositions of S^2 induced by f and $\mathscr{C}$ in this proof.

Recall that $\mathbf{W}^i$ defined in (2.6) denotes the set of all i-flowers $W^i(p)$, $p \in \mathbf{V}^i$, for each $i \in \mathbb{N}_0$.

Since f is expanding, it is easy to see from Lemma 2.13, Proposition 2.6, and the Lebesgue Number Lemma ([Mu00, Lemma 27.5]) that $\{\mathbf{W}^i\}_{i\in\mathbb{N}_0}$ forms a refining sequence of open covers of S^2 (see Definition 3.1). Thus it suffices to prove that

$$h^*(f) = \lim_{m\to+\infty} \lim_{l\to+\infty} \lim_{n\to+\infty} \frac{1}{n} H\left(\bigvee_{i=0}^{n-1} f^{-i}(\mathbf{W}^l) \middle| \bigvee_{j=0}^{n-1} f^{-j}(\mathbf{W}^m) \right) = 0. \tag{6.4}$$

See (3.14) for the definition of H.

We now fix arbitrary $n, m, l \in N$ that satisfy $m + n > l > m$.

The plan for the proof is the following. We will first obtain an upper bound for the number of $(m + n - 1)$-flowers needed to cover each element A in the cover $\bigvee_{j=0}^{n-1} f^{-j}(\mathbf{W}^m)$ of S^2. By Lemma 6.8, it suffices to find an upper bound for card $E_m(p_0, p_1, \dots, p_{n-2}; p_{n-1})$ for $p_0, p_1, \dots, p_{n-1} \in \mathbf{V}^m$. We identify $E_m(p_0, p_1, \dots, p_{n-2}; p_{n-1})$ with the set of leaves of a certain rooted tree. By Lemma 6.9, we will only need to bound the number of vertices with more than one child in each path connecting the root with some leave. This can be achieved after one observes that for an expanding Thurston map with no periodic critical points, the frequency for an orbit getting near the set of critical points is bounded from above. After this main step, we will then find an upper bound for the number of $(l+n)$-tiles needed to cover A. By observing that each $(l + n)$-tile is a subset of some element in $\bigvee_{j=0}^{n-1} f^{-j}(\mathbf{W}^l)$, we will finally obtain a suitable upper bound for $H\left(\bigvee_{i=0}^{n-1} f^{-i}(\mathbf{W}^l) \middle| \bigvee_{j=0}^{n-1} f^{-j}(\mathbf{W}^m) \right)$ which leads to (6.4).

Let $A \in \bigvee_{j=0}^{n-1} f^{-j}(\mathbf{W}^m)$, say

$$A = \bigcap_{i=0}^{n-1} f^{-i}(W^m(p_i)) \tag{6.5}$$

where $p_0, p_1, \dots, p_{n-1} \in \mathbf{V}^m$. By Lemma 6.8,

$$A \subseteq \bigcup_{x \in E_m(p_0, p_1, \ldots, p_{n-2}; p_{n-1})} W^{m+n-1}(x), \tag{6.6}$$

where E_m is defined in (6.2).

We can construct a rooted tree $\mathscr{T}$ from $E_m(p_0, p_1, \ldots, p_{n-2}; p_{n-1})$ as a simple directed graph. The set $\mathscr{V}(\mathscr{T})$ of vertices of $\mathscr{T}$ is

$$\mathscr{V}(\mathscr{T}) = \bigcup_{i=0}^{n-1} \left\{ (f^i(x), n-1-i) \in S^2 \times \mathbb{N}_0 \,\middle|\, x \in E_m(p_0, p_1, \ldots, p_{n-2}; p_{n-1}) \right\}.$$

Two vertices $(x, i), (y, j) \in \mathscr{V}(\mathscr{T})$ are connected by a directed edge $((x, i), (y, j)) \in \mathscr{E}(\mathscr{T})$ if and only if $f(y) = x$ and $j = i + 1$. Clearly the simple directed graph $\mathscr{T}$ constructed this way is a finite rooted tree with root $(p_{n-1}, 0) \in \mathscr{V}(\mathscr{T})$.

Observe that if a vertex $(x, i) \in \mathscr{V}(\mathscr{T})$ is a leaf of $\mathscr{T}$, then $x \in f^{-n+1}(p_{n-1})$ and $i = n - 1$.

For each $(x, i) \in \mathscr{V}(\mathscr{T})$, we write $c(x, i) = d^+((x, i))$, i.e.,

$$c(x, i) = \operatorname{card}\{(y, i+1) \in \mathscr{V}(\mathscr{T}) \mid f(y) = x\}. \tag{6.7}$$

We make the convention that for each $x \in S^2$ and each $i \in \mathbb{Z}$, if $(x, i) \notin \mathscr{V}(\mathscr{T})$, then $c(x, i) = -1$. See Fig. 6.2 for an example of $\mathscr{T}$.

Recall that by (6.2),

$$\begin{aligned} &E_m(p_0, p_1, \ldots, p_{n-2}; p_{n-1}) \\ =&\left\{ y \in f^{-n+1}(p_{n-1}) \,\middle|\, f^i(y) \in \overline{W}^m(p_i), i \in \{0, 1, \ldots, n-2\} \right\}. \end{aligned}$$

So if $(x, i) \in \mathscr{V}(\mathscr{T})$, then $c(x, i)$ is at most the number of distinct preimages of x under f contained in $\overline{W}^m(p_{i+1})$. Thus

$$0 \leq c(x, i) \leq \deg f \text{ for } (x, i) \in \mathscr{V}(\mathscr{T}). \tag{6.8}$$

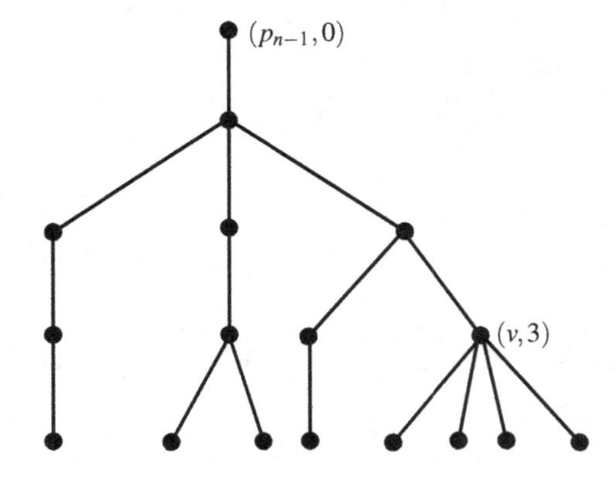

Fig. 6.2 An example of $\mathscr{T}$ with $n = 5$ and $c(v, 3) = 4$

Fix a visual metric d on S^2 for f with expansion factor $\Lambda > 1$. The map f is Lipschitz with respect to d (see Lemma 2.15). Then there exists a constant $K \geq 1$ depending only on f and d such that $d(f(x), f(y)) \leq Kd(x, y)$ for $x, y \in S^2$. We may assume that $K \geq 2$.

Define

$$N_c = \max\{\min\{i \in \mathbb{N} \mid f^j(x) \notin \text{crit } f \text{ if } j \geq i\} \mid x \in \text{crit } f\}.$$

The maximum is taken over a finite set of integers since f has no periodic critical points. So $N_c \in \mathbb{N}$. Note that by definition, if $x \in \text{crit } f$, then $f^i(x) \in \text{post } f \setminus \text{crit } f$ for each $i \geq N_c$. Denote the shortest distance between a critical point and the set post $f \setminus$ crit f by

$$D_c = \min\{d(x, y) \mid x \in \text{post } f \setminus \text{crit } f, y \in \text{crit } f\}.$$

Then $D_c \in (0, +\infty)$ since both post $f \setminus$ crit f and crit f are nonempty finite sets.

We now proceed to find an upper bound for

$$\text{card } \big\{i \in \{0, 1, \ldots, n-1\} \,\big|\, c(f^i(z), n-1-i) \geq 2\big\}$$

for each $(z, n-1) \in \mathcal{V}(\mathcal{T})$, uniform in $(z, n-1)$. Recall that $z \in f^{-n+1}(p_{n-1})$ for each $(z, n-1) \in \mathcal{V}(\mathcal{T})$. We fix such a point z.

In order to find an upper bound, we first define, for each $i \in \mathbb{N}$ sufficiently large,

$$M_i = \left\lfloor \log_K \left(\frac{D_c - \tau(3C\Lambda^{-i})}{\tau(3C\Lambda^{-i})} \right) \right\rfloor - 2, \tag{6.9}$$

where the function τ is from Lemma 6.6, and $C \geq 1$ is a constant depending only on f, $\mathcal{C}$, and d from Lemma 2.13. Note that $\tau(3C\Lambda^{-i}) \longrightarrow 0$ as $i \longrightarrow +\infty$ (Lemma 6.6), thus M_i is well-defined for i sufficiently large, and

$$\lim_{i \to +\infty} M_i = +\infty. \tag{6.10}$$

We assume that m is sufficiently large such that the following conditions are both satisfied:

(i) $m > \log_\Lambda \left(\frac{3C}{\delta_0}\right)$,
(ii) $M_m > N_c$,

where $\delta_0 \in (0, 1]$ is a constant that depends only on f and d from Lemma 6.6. Note that by Lemma 2.13, each m-flower is of diameter at most $2C\Lambda^{-m}$. Thus condition (i) implies that for each $v \in \mathbf{V}^m$, each pair of points $x, y \in \overline{W}^m(v)$ satisfy $d(x, y) < 3C\Lambda^{-m} < \delta_0$.

Fix $k \in \{0, 1, \ldots, n-1\}$ with $c\big(f^k(z), n-1-k\big) \geq 2$. Then $k \neq 0$ and the number of distinct points in $\overline{W}^m(p_{k-1})$ that are mapped to $f^k(z)$ under f is at least

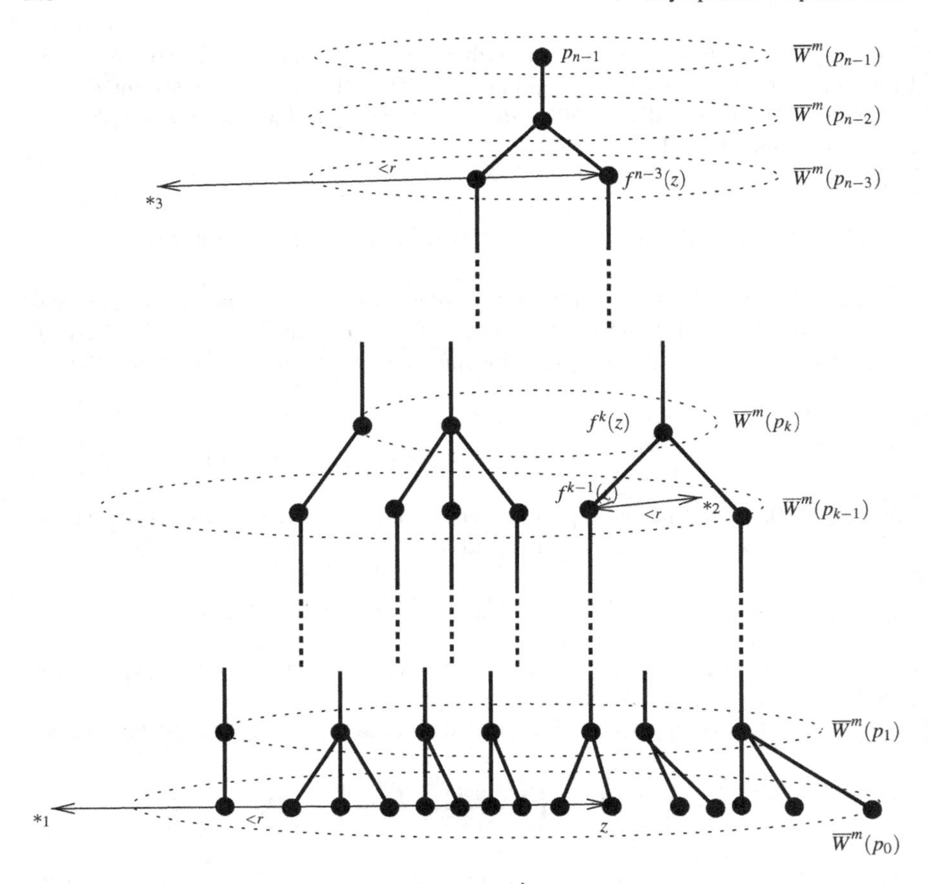

Fig. 6.3 $*_1, *_2, *_3 \in \text{crit } f$, $r = \tau(3C\Lambda^{-m})$, and $c(f^k(z), n-1-i) = 2$

$c\big(f^k(z), n-1-k\big) \geq 2$. Thus f is not injective on $\overline{W}^m(p_{k-1})$. See Fig. 6.3. By Lemma 2.13, $\text{diam}_d\left(\overline{W}^m(p_{k-1})\right) \leq 2C\Lambda^{-m}$. Since $f^{k-1}(z) \in \overline{W}^m(p_{k-1})$, the map f is not injective on $B_d\big(f^{k-1}(z), 3C\Lambda^{-m}\big)$. Then since $3C\Lambda^{-m} < \delta_0$, by Lemma 6.6,

$$d\left(f^{k-1}(z), \text{crit } f\right) < \tau\big(3C\Lambda^{-m}\big).$$

Choose $w \in \text{crit } f$ that satisfies $d\left(f^{k-1}(z), w\right) < \tau\big(3C\Lambda^{-m}\big)$. Then for each $j \in \mathbb{N}_0$,

$$d\left(f^{k+j-1}(z), f^j(w)\right) < K^j \tau\big(3C\Lambda^{-m}\big). \tag{6.11}$$

We will show that in the sequence $f^k(z), f^{k+1}(z), \ldots, f^{k+M_m}(z)$, the number of terms $f^{k+j}(z)$, $0 \leq j \leq M_m$, for which the vertex

$$\left(f^{k+j}(z), n-1-k-j\right) \in \mathscr{V}(\mathscr{T})$$

has at least two children is bounded above by N_c, i.e.,

$$\operatorname{card}\left\{j \in \{0, 1, \dots, M_m\} \,\middle|\, c\left(f^{k+j}(z), n-1-k-j\right) \geq 2\right\} \leq N_c. \tag{6.12}$$

Note that M_m is defined in (6.9). Here we use the convention that for each $x \in S^2$ and each $i \in \mathbb{Z}$, if $(x, i) \notin \mathscr{V}(\mathscr{T})$, then $c(x, i) = -1$.

Indeed, for each $j \in \{N_c, N_c + 1, \dots, \min\{M_m, n-1-k\}\}$, we have $f^j(w) \in \operatorname{post} f \setminus \operatorname{crit} f$. Note that here $M_m > N_c$ by condition (ii) on m. Thus by (6.11) and (6.9),

$$\begin{aligned} d\left(f^{k+j-1}(z), \operatorname{crit} f\right) &\geq d\left(\operatorname{crit} f, f^j(w)\right) - d\left(f^j(w), f^{k+j-1}(z)\right) \\ &\geq D_c - K^j \tau(3C\Lambda^{-m}) \\ &\geq D_c - K^{M_m} \tau(3C\Lambda^{-m}) \\ &\geq D_c - \left(\frac{D_c - \tau(3C\Lambda^{-m})}{\tau(3C\Lambda^{-m})}\right) \tau(3C\Lambda^{-m}) \\ &= \tau(3C\Lambda^{-m}). \end{aligned}$$

Hence by Lemma 6.6, the restriction of f to $B_d\left(f^{k+j-1}(z), 3C\Lambda^{-m}\right)$ is injective. Note that $f^{k+j-1}(z) \in \overline{W}^m(p_{k+j-1})$, and by Lemma 2.13, $\operatorname{diam}_d\left(\overline{W}^m(p_{k+j-1})\right) \leq 2C\Lambda^{-m}$. So f is injective on $\overline{W}^m(p_{k+j-1})$. Thus

$$c\left(f^{k+j}(z), n-1-k-j\right) = 1$$

for each $j \in \{N_c, N_c + 1, \dots, \min\{M_m, n-1-k\}\}$. Hence

$$c\left(f^{k+j}(z), n-1-i-j\right) \in \{1, -1\}$$

for each $j \in \{N_c, N_c + 1, \dots, M_m\}$. Then (6.12) holds.

Thus we get that

$$\operatorname{card}\left\{i \in \{0, 1, \dots, n-1\} \,\middle|\, c(f^i(z), n-1-i) \geq 2\right\} \leq N_c \left\lceil \frac{n}{M_m} \right\rceil \tag{6.13}$$

for each $(z, n-1) \in \mathscr{V}(\mathscr{T})$.

Hence by (6.13), (6.8), and Lemma 6.9, we can conclude that the number of leaves of $\mathscr{T}$ is at most $(\deg f)^{N_c\left(\frac{n}{M_m}+1\right)}$, or equivalently,

$$\operatorname{card} E_m(p_0, p_1, \dots, p_{n-2}; p_{n-1}) \leq (\deg f)^{N_c\left(\frac{n}{M_m}+1\right)}. \tag{6.14}$$

We have obtained an upper bound for the number of $(m+n-1)$-flowers needed to cover A. Next, we will find an upper bound for the number of $(m+n-1)$-tiles, and consequently, an upper bound for the number of $(l+n)$-tiles, needed to cover A.

Denote the maximum number of i-tiles contained in the closure of any i-flower, over all $i \in \mathbb{N}_0$, by W_f, i.e.,

$$W_f = \sup\left\{ \operatorname{card}\left\{X^i \in \mathbf{X}^i \mid X^i \subseteq \overline{W}^j(v)\right\} \mid j \in \mathbb{N}_0, v \in \mathbf{V}^j\right\}.$$

Observe that $W_f = \sup\{2 \deg_{f^i}(v) \mid i \in \mathbb{N}_0,\ v \in \mathbf{V}^i\}$. Since f has no periodic critical points, it follows from [BM17, Lemma 18.6] that W_f is a finite number that only depends on f.

Thus we can cover A in (6.5) by a collection of $(m + n - 1)$-tiles of cardinality at most $W_f(\deg f)^{N_c\left(\frac{n}{M_m}+1\right)}$.

On the other hand, we claim that each $(l + n)$-tile $X^{l+n} \in \mathbf{X}^{l+n}$ is a subset of at least one element in the open cover $\bigvee_{i=0}^{n-1} f^{-i}(\mathbf{W}^l)$ of S^2. To prove the claim, we first fix an $(l + n)$-tile $X^{l+n} \in \mathbf{X}^{l+n}$. By Proposition 2.6(ii) and Lemma 6.7(ii), for each $i \in \{0, 1, \dots, n - 1\}$, there exists an l-vertex $v_i \in \mathbf{V}^l$ such that $f^i\left(X^{l+n}\right) \subseteq W^l(v_i)$. Thus

$$X^{l+n} \subseteq \bigcap_{i=0}^{n-1} f^{-i}\left(W^l(v_i)\right).$$

The proof for the claim is complete.

Note that for each $(m + n - 1)$-tile $X^{m+n-1} \in \mathbf{X}^{m+n-1}$, the collection

$$\left\{X^{l+n} \in \mathbf{X}^{l+n} \mid X^{l+n} \subseteq X^{m+n-1}\right\}$$

forms a cover of X^{m+n-1}, and has cardinality at most $(2 \deg f)^{l-m+1}$, which follows immediately from Proposition 2.6.

Hence, we get that for each element A of $\bigvee_{j=0}^{n-1} f^{-j}(\mathbf{W}^m)$, we can find a cover of A consisting of elements of $\bigvee_{i=0}^{n-1} f^{-i}(\mathbf{W}^l)$ in such a way that the cardinality of the cover is at most $(2 \deg f)^{l-m+1} W_f(\deg f)^{N_c\left(\frac{n}{M_m}+1\right)}$.

We conclude that

$$\begin{aligned}
h^*(f) &\leq \lim_{m\to+\infty} \lim_{l\to+\infty} \lim_{n\to+\infty} \frac{1}{n} \log\left((2\deg f)^{l-m+1} W_f(\deg f)^{N_c\left(\frac{n}{M_m}+1\right)}\right) \\
&= \lim_{m\to+\infty} \lim_{l\to+\infty} \lim_{n\to+\infty} \frac{1}{n} N_c\left(\frac{n}{M_m}+1\right) \log(\deg f) \\
&= \lim_{m\to+\infty} \frac{N_c \log(\deg f)}{M_m} \\
&= 0.
\end{aligned}$$

The last equality follows from (6.10). Therefore $h^*(f) = 0$. □

Recall that a point $x \in S^2$ is a periodic point of $f: S^2 \to S^2$ with period n if $f^n(x) = x$ and $f^i(x) \neq x$ for each $i \in \{1, 2, \dots, n-1\}$.

Theorem 6.11 *An expanding Thurston map $f: S^2 \to S^2$ with at least one periodic critical point is not asymptotically h-expansive.*

Proof We need to show $h^*(f) > 0$. By (3.16), it suffices to prove that f^i is not asymptotically h-expansive for some $i \in \mathbb{N}$. Note that by (2.2), if a point $x \in S^2$ is a periodic critical point of f^i for some $i \in \mathbb{N}$, then it is a periodic point of f and there exists $j \in \mathbb{N}_0$ such that $f^j(x)$ is a periodic critical point of f. Thus each periodic critical point of f^τ is a fixed point of f^τ if $\tau \in \mathbb{N}$ is a common multiple of the periods of all the periodic critical points of f. Hence by Lemma 2.17, we can assume, without loss of generality, that there exists a Jordan curve $\mathscr{C} \subseteq S^2$ containing post f such that $f(\mathscr{C}) \subseteq \mathscr{C}$, and no 1-tile joins opposite sides of $\mathscr{C}$, and each periodic critical point of f is a fixed point of f.

Let p be a critical point of f that is fixed by f.

In addition, we can assume, without loss of generality, that $f^{-1}(p) \setminus \mathscr{C} \neq \emptyset$. Indeed, by Lemma 2.12, there exists $j \in \mathbb{N}$ such that $f^{-j}(p) \setminus \mathscr{C} \neq \emptyset$. We replace f by f^j, and observe that by (2.2) and the fact that each periodic critical point of f is a fixed point of f, the set of periodic critical points of f and that of f^j coincide. Note that for the new map and its invariant curve $\mathscr{C}$, no 1-tile joins opposite sides of $\mathscr{C}$, and each periodic critical point is a fixed point.

From now on, we consider the cell decompositions of S^2 induced by f and $\mathscr{C}$ in this proof.

Recall that for $i \in \mathbb{N}_0$, we denote by $\mathbf{W}^i$ as in (2.6) the set of all i-flowers $W^i(p)$ where $p \in \mathbf{V}^i$.

Since f is expanding, it is easy to see from Lemma 2.13, Proposition 2.6, and the Lebesgue Number Lemma ([Mu00, Lemma 27.5]) that $\{\mathbf{W}^i\}_{i \in \mathbb{N}_0}$ forms a refining sequence of open covers of S^2 (see Definition 3.1). Thus it suffices to prove that

$$h^*(f) = \lim_{m \to +\infty} \lim_{l \to +\infty} \lim_{n \to +\infty} \frac{1}{n} H\left(\bigvee_{i=0}^{n-1} f^{-i}(\mathbf{W}^l) \middle| \bigvee_{j=0}^{n-1} f^{-j}(\mathbf{W}^m) \right) > 0.$$

See (3.14) for the definition of H.

Our plan is to construct a sequence $\{v_i\}_{i \in \mathbb{N}}$ of m-vertices in such a way that for each $n \in \mathbb{N}$, the number of elements in $\bigvee_{i=0}^{n-1} f^{-i}(\mathbf{W}^l)$ needed to cover $B_n = \bigcap_{j=0}^{n-1} f^{-j}(W^m(v_{n-j}))$ can be bounded from below in such a way that $h^*(f) > 0$ follows immediately. More precisely, we observe that the more connected components B_n has, the harder to cover B_n. So we will choose $\{v_i\}_{i \in \mathbb{N}}$ as a periodic sequence of m-vertices shadowing an infinite backward pseudo-orbit under iterations of f in such a way that each period of $\{v_i\}_{i \in \mathbb{N}}$ begins with a backward orbit starting at p and approaching p as the index i increases, and then ends with a constant sequence staying at p. By a recursive construction, we keep track of each B_n by a finite subset $V_n \subseteq B_n$ with the property that $\operatorname{card}(A \cap V_n) \leq 1$ for each $A \in \bigvee_{i=0}^{n-1} f^{-i}(\mathbf{W}^l)$. A quantitative control of the size of V_n leads to the conclusion that $h^*(f) > 0$. The fact that the constant part of each period of $\{v_i\}_{i\in\mathbb{N}}$ can be made arbitrarily long is essential here and is not true if f has no periodic critical points.

For this we fix $m, l \in \mathbb{N}$ with $l > m + 100$.

Let $k = \deg_f(p)$. Then $k > 1$.

Define $q_0 = p$ and choose $q_1 \in f^{-1}(p) \setminus \mathscr{C}$. Then q_1 is necessarily a 1-vertex, but not a 0-vertex, i.e., $q_1 \in \mathbf{V}^1 \setminus \mathbf{V}^0$. Since $q_1 \notin \mathscr{C}$, we have $q_1 \in W^0(p)$. By (2.5), the only 2-vertex contained in $W^2(p)$ is p. So $q_1 \in W^0(p) \setminus W^2(p)$. Since $f\big(W^i(p)\big) = W^{i-1}(p)$ for each $i \in \mathbb{N}$ (see Remark 2.7), we can recursively choose $q_j \in \mathbf{V}^j$ for $j \in \{2, 3, \dots, m\}$ such that

(i) $f(q_j) = q_{j-1}$,
(ii) $q_j \in W^{j-1}(p) \setminus W^{j+1}(p)$.

We define a singleton set $Q_m = \{q_m\}$.

We set $q_m^1 = q_m$.

Next, we choose recursively, for each $j \in \{m + 1, m + 2, \dots, l - 2\}$, a set Q_j with $\operatorname{card} Q_j = k^{j-m}$ consisting of distinct points $q_j^i \in \mathbf{V}^j$, $i \in \{1, 2, \dots, k^{j-m}\}$, such that

(i) $f(Q_j) = Q_{j-1}$,
(ii) $Q_j \subseteq W^{j-1}(p) \setminus W^{j+1}(p)$.

Note by Remark 2.7, it is clear that these two properties uniquely determines Q_j from Q_{j-1}.

Finally, we construct recursively, for $j \in \{l-1, l, l+1\}$, a set Q_j with $\operatorname{card} Q_j = k^{l-2-m}$ consisting of distinct points $q_j^i \in \mathbf{V}^j$, $i \in \{1, 2, \dots, k^{l-2-m}\}$, such that

(i) $f(q_j^i) = q_{j-1}^i$,
(ii) $Q_j \subseteq W^{j-1}(p) \setminus W^{j+1}(p)$.

We will now construct recursively, for each $n = (l + 1)s + r$, with $s \in \mathbb{N}_0$ and $r \in \{0, 1, \dots, l\}$, an m-vertex $v_n \in \mathbf{V}^m$ and a set of n-vertices $V_n \subseteq \mathbf{V}^n$ such that the following properties are satisfied:

(1) $V_n \subseteq W^m(v_n)$ for $n \in \mathbb{N}_0$;
(2) $f\big(V_n\big) = V_{n-1}$ for $n \in \mathbb{N}$;
(3) For $s \in \mathbb{N}_0$, and
 (i) for $r = 0$, $V_{(l+1)s+r} \subseteq W^l(p)$,
 (ii) for $r \in \{1, 2, \dots, m\}$, $V_{(l+1)s+r} \subseteq W^{l+1}\big(v_{(l+1)s+r}\big)$,
 (iii) for $r \in \{m+1, m+2, \dots, l-2\}$, there exists, for each $i \in \{1, 2, \dots, k^{r-m}\}$, a subset $V^i_{(l+1)s+r}$ of $V_{(l+1)s+r}$ such that
  (a) $V^i_{(l+1)s+r} \cap V^j_{(l+1)s+r} = \emptyset$ for $1 \leq i < j \leq k^{r-m}$,

(b) $\bigcup_{i=1}^{k^{r-m}} V^i_{(l+1)s+r} = V_{(l+1)s+r}$,

(c) $V^i_{(l+1)s+r} \subseteq W^{l+1}(q^i_r)$,

(iv) for $r \in \{l-1, l\}$, there exists, for each $i \in \{1, 2, \ldots, k^{l-2-m}\}$, a subset $V^i_{(l+1)s+r}$ of $V_{(l+1)s+r}$ such that

(a) $V^i_{(l+1)s+r} \cap V^j_{(l+1)s+r} = \emptyset$ for $1 \le i < j \le k^{l-2-m}$,

(b) $\bigcup_{i=1}^{k^{l-2-m}} V^i_{(l+1)s+r} = V_{(l+1)s+r}$,

(c) $V^i_{(l+1)s+r} \subseteq W^{l+1}(q^i_r)$;

(4) for $n \in \mathbb{N}_0$, $A \in \bigvee_{i=0}^{n-1} f^{-i}(\mathbf{W}^l)$, and $x, y \in V_n$ with $x \neq y$, we have $\{x, y\} \not\subseteq A$.

We start our construction by first defining $v_n \in \mathbf{V}^m$ for each $n \in \mathbb{N}$. For $s \in \mathbb{N}_0$ and $r \in \{0, 1, \ldots, m\}$, set $v_{(l+1)s+r} = q_r$. For $s \in \mathbb{N}_0$ and $r \in \{m+1, m+2, \ldots, l\}$, set $v_{(l+1)s+r} = p$.

We now define V_n recursively.

Let $V_0 = \{q_0\}$. Clearly V_0 satisfies properties (1) through (4).

Assume that V_n is defined and satisfies properties (1) through (4) for each $n \in \{0, 1, \ldots, (l+1)s+r\}$, where $s \in \mathbb{N}_0$ and $r \in \{0, 1, \ldots, l\}$. We continue our construction in the following cases depending on r.

Case 1. Assume $r \in \{0, 1, \ldots, m-1\}$. Then $v_{(l+1)s+r} = q_r$ and $v_{(l+1)s+r+1} = q_{r+1}$.

Since $f(W^{l+1}(q_{r+1})) = W^l(q_r)$ (see Remark 2.7), and $V_{(l+1)s+r} \subseteq W^l(q_r)$ by the induction hypothesis, we can choose, for each $x \in V_{(l+1)s+r}$, a point $x' \in W^{l+1}(q_{r+1})$ such that $f(x') = x$. Then define $V_{(l+1)s+r+1}$ to be the collection of all such chosen x' that corresponds to $x \in V_{(l+1)s+r}$. Note that

$$\operatorname{card} V_{(l+1)s+r+1} = \operatorname{card} V_{(l+1)s+r}.$$

All properties required for $V_{(l+1)s+r+1}$ in the induction step are trivial to verify. We only consider the last property here. Indeed, suppose that $x, y \in V_{(l+1)s+r+1}$ satisfy that $x \neq y$ and $\{x, y\} \subseteq A$ for some $A \in \bigvee_{i=0}^{(l+1)s+r} f^{-i}(\mathbf{W}^l)$. Then by construction $f(x)$, $f(y)$, and $f(A)$ satisfy

(a) $f(A) \subseteq B$ for some $B \in \bigvee_{i=0}^{(l+1)s+r-1} f^{-i}(\mathbf{W}^l)$,

(b) $f(x), f(y) \in V_{(l+1)s+r}$, and $f(x) \neq f(y)$,

(c) $\{f(x), f(y)\} \subseteq f(A) \subseteq B$.

This contradicts property (4) for $V_{(l+1)s+r}$ in the induction hypothesis.

Case 2. Assume $r \in \{m, m+1, \ldots, l-3\}$. Then $v_{(l+1)s+r+1} = p$, $v_{(l+1)s+m} = q_m$, and when $r \neq m$, we have $v_{(l+1)s+r} = p$.

If $r = m$, we define $V^1_{(l+1)s+r} = V_{(l+1)s+r}$. Recall that $q^1_m = q_m$.

Note that for each $i \in \{1, 2, \ldots, k^{r+1-m}\}$, $f\big(W^{l+2}(q^i_{r+1})\big) = W^{l+1}\big(q^j_r\big)$ for some $j \in \{1, 2, \ldots, k^{r-m}\}$ (see Remark 2.7), and $V^j_{(l+1)s+r} \subseteq W^{l+1}\big(q^j_r\big)$ by the induction hypothesis. For each $j \in \{1, 2, \ldots, k^{r-m}\}$, each $x \in V^j_{(l+1)s+r}$, and each $i \in \{1, 2, \ldots, k^{r+1-m}\}$ with $f\big(W^{l+2}(q^i_{r+1})\big) = W^{l+1}\big(q^j_r\big)$, we can choose a point $x' \in W^{l+2}\big(q^i_{r+1}\big)$ such that $f(x') = x$. Then define $V^i_{(l+1)s+r+1}$ to be the collection of all such chosen x' that corresponds to $x \in V^j_{(l+1)s+r}$. Set $V_{(l+1)s+r+1} = \bigcup_{i=1}^{k^{r+1-m}} V^i_{(l+1)s+r+1}$.

Since $Q_{r+1} \subseteq \mathbf{V}^{r+1} \cap W^m(p)$, $r \in \{m, m+1, \ldots, l-3\}$, $l > m + 100$, and no 1-tile joins opposite sides of $\mathscr{C}$, we get that

(a) for $i, j \in \{1, 2, \ldots, k^{r+1-m}\}$ with $i \neq j$, by Lemma 6.7(iii),

$$W^{l+2}\big(q^i_{r+1}\big) \cap W^{l+2}\big(q^j_{r+1}\big) = \emptyset,$$

and so $V^i_{(l+1)s+r+1} \cap V^j_{(l+1)s+r+1} = \emptyset$,

(b) $V_{(l+1)s+r+1} \subseteq W^m(p)$.

Thus

$$\text{card } V_{(l+1)s+r+1} = k \text{ card } V_{(l+1)s+r}.$$

We only need to verify property (4) required for $V_{(l+1)s+r+1}$ in the induction step now. Indeed, suppose that $x, y \in V_{(l+1)s+r+1}$ with $x \neq y$ and $\{x, y\} \subseteq A$ for some $A \in \bigvee_{a=0}^{(l+1)s+r} f^{-a}\big(\mathbf{W}^l\big)$. Then $A \subseteq W^l(v^l)$ for some $v^l \in \mathbf{V}^l$. By construction, there exist $i, j \in \{1, 2, \ldots, k^{r+1-m}\}$ such that $x \in W^{l+2}\big(q^i_{r+1}\big)$ and $y \in W^{l+2}\big(q^j_{r+1}\big)$. Note that $q^i_{r+1}, q^j_{r+1} \in \mathbf{V}^{r+1}$, $r \in \{m, m+1, \ldots, l-3\}$, and $l > m+100$. So $q^i_{r+1}, q^j_{r+1} \in \mathbf{V}^{l-2}$. Since $x \in W^l(v^l) \cap W^{l+2}\big(q^i_{r+1}\big)$, we get $q^i_{r+1} \in \overline{W}^l(v_l)$ by Lemma 6.7(i), and thus $v^l \in \overline{W}^l(q^i_{r+1})$. Similarly $v^l \in \overline{W}^l(q^j_{r+1})$. Since $q^i_{r+1}, q^j_{r+1} \in \mathbf{V}^{l-2}$ and no 1-tile joins opposite sides of $\mathscr{C}$, we get from Lemma 6.7(iii) that $q^i_{r+1} = q^j_{r+1}$, i.e., $i = j$. Thus $f(x) \neq f(y)$ by construction. But then $f(x)$, $f(y)$, and $f(A)$ satisfy

(a) $f(A) \subseteq B$ for some $B \in \bigvee_{a=0}^{(l+1)s+r-1} f^{-a}\big(\mathbf{W}^l\big)$,

(b) $f(x), f(y) \in V_{(l+1)s+r}$, and $f(x) \neq f(y)$,

(c) $\{f(x), f(y)\} \subseteq f(A) \subseteq B$.

This contradicts property (4) for $V_{(l+1)s+r}$ in the induction hypothesis.

Case 3. Assume $r \in \{l-2, l-1, l\}$, then $v_{(l+1)s+r+1} = v_{(l+1)s+r} = p$.

Note that for each $i \in \{1, 2, \ldots, k^{l-2-m}\}$, $f\big(W^{l+2}(q^i_{r+1})\big) = W^{l+1}\big(q^i_r\big)$ (see Remark 2.7), and $V^i_{(l+1)s+r} \subseteq W^{l+1}\big(q^i_r\big)$ by the induction hypothesis. For each $j \in \{1, 2, \ldots, k^{l-2-m}\}$ and each $x \in V^i_{(l+1)s+r}$, we can choose a point $x' \in W^{l+2}\big(q^i_{r+1}\big)$ such that $f(x') = x$. Then define $V^i_{(l+1)s+r+1}$ to be the collection of all such chosen x' that corresponds to $x \in V^i_{(l+1)s+r}$. Set $V_{(l+1)s+r+1} = \bigcup_{i=1}^{k^{l-2-m}} V^i_{(l+1)s+r+1}$.

Since $Q_{r+1} \subseteq \mathbf{V}^{r+1} \cap W^r(p)$, $r \in \{l-2, l-1, l\}$, and $l > m + 100$, we get that

(a) for $i, j \in \{1, 2, \ldots, k^{l-2-m}\}$ with $i \neq j$,

$$f\left(V^i_{(l+1)s+r+1}\right) \cap f\left(V^j_{(l+1)s+r+1}\right) = V^i_{(l+1)s+r} \cap V^j_{(l+1)s+r} = \emptyset$$

(by the induction hypothesis), and so

$$V^i_{(l+1)s+r+1} \cap V^j_{(l+1)s+r+1} = \emptyset,$$

(b) $V_{(l+1)s+r+1} \subseteq W^m(p)$,
(c) if $r = l$, then $V_{(l+1)s+r+1} \subseteq W^l(p)$.

Thus

$$\operatorname{card} V_{(l+1)s+r+1} = \operatorname{card} V_{(l+1)s+r}.$$

We only need to verify the last property required for $V_{(l+1)s+r+1}$ in the induction step now. Indeed, suppose that $x, y \in V_{(l+1)s+r+1}$ with $x \neq y$ and $\{x, y\} \subseteq A$ for some $A \in \bigvee_{i=0}^{(l+1)s+r} f^{-i}(\mathbf{W}^l)$. Then by construction $f(x)$, $f(y)$, and $f(A)$ satisfy

(a) $f(A) \subseteq B$ for some $B \in \bigvee_{i=0}^{(l+1)s+r-1} f^{-i}(\mathbf{W}^l)$,
(b) $f(x), f(y) \in V_{(l+1)s+r}$, and $f(x) \neq f(y)$,
(c) $\{f(x), f(y)\} \subseteq f(A) \subseteq B$.

This contradicts property (4) for $V_{(l+1)s+r}$ in the induction hypothesis.

The recursive construction and the inductive proof of the properties of the construction are now complete.

Note that by our construction, we have

$$\operatorname{card} V_{(l+1)s} = k^{(l-m-2)s}, \qquad s \in \mathbb{N}. \tag{6.15}$$

For each $s \in \mathbb{N}$, we consider

$$B_{(l+1)s} = \bigcap_{j=0}^{(l+1)s-1} f^{-j}\left(W^m\left(v_{(l+1)s-j}\right)\right) \in \bigvee_{j=0}^{(l+1)s-1} f^{-j}(\mathbf{W}^m).$$

Then $V_{(l+1)s} \subseteq B_{(l+1)s}$ by properties (1) and (2) of the construction. On the other hand, by property (4), if $\mathscr{A} \subseteq \bigvee_{j=0}^{(l+1)s-1} f^{-j}(\mathbf{W}^l)$ satisfies

$$\bigcup \mathscr{A} \supseteq B_{(l+1)s} \supseteq V_{(l+1)s}.$$

So $\operatorname{card} \mathscr{A} \geq \operatorname{card} V_{(l+1)s}$.

Thus by (3.15), (3.14), and (6.15),

$$\begin{aligned}
h^*(f) &= \lim_{m\to+\infty}\lim_{l\to+\infty}\lim_{n\to+\infty}\frac{1}{n}H\left(\bigvee_{i=0}^{n-1} f^{-i}(\mathbf{W}^l)\middle|\bigvee_{j=0}^{n-1} f^{-j}(\mathbf{W}^m)\right)\\
&\geq \liminf_{m\to+\infty}\liminf_{l\to+\infty}\liminf_{s\to+\infty}\frac{1}{(l+1)s}\log\left(k^{(l-m-2)s}\right)\\
&= \liminf_{m\to+\infty}\liminf_{l\to+\infty}\frac{l-m-2}{l+1}\log k\\
&= \log k\\
&> 0.
\end{aligned}$$

Therefore, the map f is not asymptotically h-expansive. □

Lemma 6.12 *Let $g : X \to X$ be a continuous map on a compact metric space (X, d). If g is h-expansive then so is g^n for each $n \in \mathbb{N}$.*

The converse can also be easily established, i.e., if g^n is h-expansive for some $n \in \mathbb{N}$, then so is g. But we will not need it in this paper.

Proof We first observe from Definition 3.1 that if $\{\xi_l\}_{l\in\mathbb{N}_0}$ is a refining sequence of open covers, then so is $\{\xi_l^n\}_{l\in\mathbb{N}_0}$ for each $n \in \mathbb{N}$, where $\xi_l^n = \bigvee_{i=0}^{n-1} g^{-i}(\xi_l)$. We also note that given an open cover λ of X, we have

$$\bigvee_{i=0}^{mn-1} g^{-i}(\lambda) = \bigvee_{j=0}^{m-1} (g^n)^{-j}(\lambda^n)$$

for $n, m \in \mathbb{N}$, where $\lambda^n = \bigvee_{k=0}^{n-1} g^{-k}(\lambda)$.

Assume that g is h-expansive, then $h(g|\lambda) = 0$ for some finite open cover λ of X. Thus for each $n \in \mathbb{N}$,

$$\begin{aligned}
h(g|\lambda) &= \lim_{l\to+\infty}\lim_{m\to+\infty}\frac{1}{mn}H\left(\bigvee_{i=0}^{mn-1} g^{-i}(\xi_l)\middle|\bigvee_{j=0}^{mn-1} g^{-j}(\lambda)\right)\\
&=\frac{1}{n}\lim_{l\to+\infty}\lim_{m\to+\infty}\frac{1}{m}H\left(\bigvee_{i=0}^{m-1} (g^n)^{-i}(\xi_l^n)\middle|\bigvee_{j=0}^{m-1} (g^n)^{-j}(\lambda^n)\right)\\
&=\frac{1}{n}h\left(g^n|\lambda^n\right),
\end{aligned}$$

where ξ_l^n, λ^n are defined as above. Note that λ^n is also a finite open cover of X. Therefore $h\left(g^n|\lambda^n\right) = 0$, i.e., g^n is h-expansive. □

The proof of the following theorem is similar to that of Theorem 6.11, and slightly simpler. However, due to subtle differences in both notation and constructions, we include the proof for the convenience of the reader.

Theorem 6.13 *No expanding Thurston map is h-expansive.*

Proof Let f be an expanding Thurston map.

By Theorem 6.11 and the fact that if f is h-expanding then it is asymptotically h-expansive (see [Mis76, Corollary 2.1]), we can assume that f has no periodic critical points.

Note that by (2.2), if a point $x \in S^2$ is a periodic critical point of f^i for some $i \in \mathbb{N}$, then there exists $j \in \mathbb{N}_0$ such that $f^j(x)$ is a periodic critical point of f. So f^i has no periodic critical points for $i \in \mathbb{N}$.

By Lemma 6.12, it suffices to prove that there exists $i \in \mathbb{N}$ such that f^i is not h-expansive. Thus by Lemma 2.17, we can assume, without loss of generality, that there exists a Jordan curve $\mathscr{C} \subseteq S^2$ containing post f such that $f(\mathscr{C}) \subseteq \mathscr{C}$ and no 1-tile joins opposite sides of $\mathscr{C}$.

In addition, we can assume, without loss of generality, that there exists a critical point $p \in \operatorname{crit} f \setminus \mathscr{C}$ with $f^2(p) = f(p) \neq p$. Indeed, we can choose any critical point $p_0 \in \operatorname{crit} f$, then $f^{2i}(p_0) = f^i(p_0) \neq p_0$ for some $i \in \mathbb{N}$ since f has no periodic critical points. By Lemma 2.12, there exist $j \in \mathbb{N}$ and $p \in f^{-ij}(p_0) \setminus \mathscr{C}$. We replace f by $f^{i(j+1)}$. Note that for this new map f, we have $p \in \operatorname{crit} f \setminus \mathscr{C}$, $f^2(p) = f(p) \neq p$, $f(\mathscr{C}) \subseteq \mathscr{C}$ and no 1-tile joins opposite sides of $\mathscr{C}$.

Let $k = \deg_f(p)$. Then $k > 1$.

From now on, we consider the cell decompositions of S^2 induced by f and $\mathscr{C}$ in this proof.

Recall that $\mathbf{W}^i$ defined in (2.6) denotes the set of all i-flowers $W^i(v)$, $v \in \mathbf{V}^i$, for each $i \in \mathbb{N}_0$.

Since f is expanding, it is easy to see from Lemma 2.13, Proposition 2.6, and the Lebesgue Number Lemma ([Mu00, Lemma 27.5]) that $\{\mathbf{W}^i\}_{i\in\mathbb{N}_0}$ forms a refining sequence of open covers of S^2 (see Definition 3.1). Thus by Remark 3.4 and Definition 3.1, it suffices to prove that

$$h(f|\mathbf{W}^m) = \lim_{l\to+\infty} \lim_{n\to+\infty} \frac{1}{n} H\left(\bigvee_{i=0}^{n-1} f^{-i}(\mathbf{W}^l) \middle| \bigvee_{j=0}^{n-1} f^{-j}(\mathbf{W}^m) \right) > 0$$

for each $m \in \mathbb{N}$ sufficient large. See (3.14) for the definition of H.

Our plan is to construct a sequence $\{v_i\}_{i\in\mathbb{N}_0}$ of m-vertices in such a way that for each $n \in \mathbb{N}_0$, the number of elements in $\bigvee_{i=0}^{n-1} f^{-i}(\mathbf{W}^l)$ needed to cover $B_n = \bigcap_{j=0}^{n-1} f^{-j}(W^m(v_{n-j}))$ can be bounded from below in such a way that $h(f|\mathbf{W}^m) > 0$ follows immediately. More precisely, we observe that the more connected components B_n has, the harder to cover B_n. So we will choose $\{v_i\}_{i\in\mathbb{N}_0}$ as a periodic sequence of m-vertices shadowing an infinite backward pseudo-orbit under iterations of f in such a way that each period of $\{v_i\}_{i\in\mathbb{N}_0}$ begins with a backward orbit starting at $f(p)$ and p, and approaching $f(p)$ as the index i increases, and then ends with $f(p)$. By a recursive construction, we keep track of each B_n by a finite subset $V_n \subseteq B_n$ with the property that card $(A \cap V_n) \leq 1$ for each $A \in \bigvee_{i=0}^{n-1} f^{-i}(\mathbf{W}^l)$. A quantitative control of the size of V_n leads to the conclusion that $h(f|\mathbf{W}^m) > 0$ for each m sufficiently large.

For this we fix $m, l \in \mathbb{N}$ with $l > 2m + 100 > 200$.

Define $q_1 = p$. Then q_1 is necessarily a 1-vertex, but not a 0-vertex, i.e., $q_1 \in \mathbf{V}^1 \setminus \mathbf{V}^0$. Since $q_1 = p \notin \mathscr{C}$, we have $q_1 \in W^0(f(p))$. By (2.5), the only 2-vertex contained in $W^2(f(p))$ is $f(p)$. So $q_1 \in W^0(f(p)) \setminus W^2(f(p))$. Since $f(W^i(f(p))) = W^{i-1}(f(p))$ for each $i \in \mathbb{N}$ (see Remark 2.7), we can recursively choose $q_j \in \mathbf{V}^j$ for each $j \in \{2, 3, \dots, m+2\}$ such that

(i) $f(q_j) = q_{j-1}$,
(ii) $q_j \in W^{j-1}(f(p)) \setminus W^{j+1}(f(p))$.

Set $q_0 = q_{m+2}$.

Since $f(W^i(p)) = W^{i-1}(f(p))$ for each $i \in \mathbb{N}$, and $k = \deg_f(p) > 1$, we can choose distinct points $p_i \in \mathbf{V}^{m+3}$, $i \in \{1, 2, \dots, k\}$, such that

(i) $f(p_i) = q_{m+2}$,
(ii) $p_i \in W^{m+2}(p) \setminus W^{m+4}(p)$.

We will now construct recursively, for each $n = (m+2)s + r$ with $s \in \mathbb{N}_0$ and $r \in \{0, 1, \dots, m+1\}$, an m-vertex $v_n \in \mathbf{V}^m$ and a set of n-vertices $V_n \subseteq \mathbf{V}^n$ such that for each $n \in \mathbb{N}_0$, the following properties are satisfied:

(1) $V_n \subseteq W^m(v_n)$;
(2) $f(V_n) = V_{n-1}$ if $n \neq 0$;
(3) (i) $V_n \subseteq W^{m+1+r}(q_r)$ if $n = (m+2)s + r$ for some $s \in \mathbb{N}_0$ and some $r \in \{1, 2, \dots, m+1\}$,
(ii) $V_n \subseteq W^{m+1+m+2}(q_0)$ if $n = (m+2)s$ for some $s \in \mathbb{N}_0$;
(4) card $V_n = k^{\lceil \frac{n}{m+2} \rceil}$;
(5) for $A \in \bigvee_{i=0}^{n-1} f^{-i}(\mathbf{W}^l)$ and $x, y \in V_n$ with $x \neq y$, we have $\{x, y\} \not\subseteq A$.

We start our construction by first defining $v_n \in \mathbf{V}^m$ for each $n \in \mathbb{N}_0$. For $s \in \mathbb{N}_0$ and $r \in \{1, 2, \dots, m\}$, set $v_{(m+2)s+r} = q_r$. For $s \in \mathbb{N}_0$ and $r \in \{0, m+1\}$, set $v_{(m+2)s+r} = f(p)$.

We now define V_n recursively.

Let $V_0 = \{q_{m+2}\}$. Clearly V_0 satisfies properties (1) through (5) in the induction step.

Assume that V_n is defined and satisfies properties (1) through (5) for each $n \in \{0, 1, \dots, (m+2)s + r\}$, where $s \in \mathbb{N}_0$ and $r \in \{0, 1, \dots, m+1\}$, we continue our construction in the following cases depending on r.

Case 1. Assume $r = 0$. Then $v_{(m+2)s+r} = f(p)$ and $v_{(m+2)s+r+1} = q_1 = p$.

Note that $V_{(m+2)s+r} \subseteq W^{2m+3}(q_r)$ by the induction hypothesis, $q_r = q_{m+2} \in W^{m+1}(f(p))$, $f(p_i) = q_r$, and $f(W^{2m+4}(p_i)) = W^{2m+3}(q_r)$ for each $i \in \{1, 2, \ldots, k\}$ (see Remark 2.7). Fix an arbitrary $i \in \{1, 2, \ldots, k\}$. We can choose, for each $x \in V_{(m+2)s+r}$, a point $x' \in W^{2m+4}(p_i)$ such that $f(x') = x$. Then define $V^i_{(m+2)s+r+1}$ to be the collection of all such chosen x' that corresponds to $x \in V_{(m+2)s+r}$. Set

$$V_{(m+2)s+r+1} = \bigcup_{i=1}^{k} V^i_{(m+2)s+r+1}.$$

Since $p_i \in W^{m+2}(p)$ and $V^i_{(m+2)s+r+1} \subseteq W^{2m+4}(p_i)$, we get that $V^i_{(m+2)s+r+1} \subseteq W^{m+2}(p)$. So $V_{(m+2)s+r+1} \subseteq W^{m+2}(p) \subseteq W^m(p)$. Since $v_{(m+2)s+r+1} = q_1 = p$, properties (1) and (3) are verified. Property (2) is clear from the construction.

To establish property (4), it suffices to show that $V^i_{(m+2)s+r+1} \cap V^j_{(m+2)s+r+1} = \emptyset$ for $1 \leq i < j \leq k$. Indeed, since $V^i_{(m+2)s+r+1} \subseteq W^{2m+4}(p_i)$ and $V^j_{(m+2)s+r+1} \subseteq W^{2m+4}(p_j)$, it suffices to prove that $\overline{W}^{2m+4}(p_i) \cap \overline{W}^{2m+4}(p_j) = \emptyset$. Suppose that $\overline{W}^{2m+4}(p_i) \cap \overline{W}^{2m+4}(p_j) \neq \emptyset$, then since no 1-tile joins opposite sides of $\mathscr{C}$, and $p_i, p_j \in \mathbf{V}^{m+3}$, we get from Lemma 6.7(iii) that $p_i = p_j$, i.e., $i = j$. But $i < j$, a contradiction.

We only need to verify property (5) now. Indeed, suppose that distinct points $x, y \in V_{(m+2)s+r+1}$ satisfy $\{x, y\} \subseteq A$ for some $A \in \bigvee_{a=0}^{(m+2)s+r} f^{-a}(\mathbf{W}^l)$. Then $A \subseteq W^l(v^l)$ for some $v^l \in \mathbf{V}^l$. By construction, there exist $i, j \in \{1, 2, \ldots, k\}$ such that $x \in W^{2m+4}(p_i)$ and $y \in W^{2m+4}(p_j)$. Since $l > 2m + 100$ and $x \in W^l(v^l) \cap W^{2m+4}(p_i)$, we get $v^l \in \overline{W}^{2m+4}(p_i)$ by Lemma 6.7(i). Similarly $v^l \in \overline{W}^{2m+4}(p_j)$. Then by the argument above, we get that $p_i = p_j$, i.e., $i = j$. Thus $f(x) \neq f(y)$ by construction. But then $f(x)$, $f(y)$, and $f(A)$ satisfy

(a) $f(A) \subseteq B$ for some $B \in \bigvee_{a=0}^{(m+2)s+r-1} f^{-a}(\mathbf{W}^l)$,
(b) $f(x), f(y) \in V_{(m+2)s+r}$, and $f(x) \neq f(y)$,
(c) $\{f(x), f(y)\} \subseteq f(A) \subseteq B$.

This contradicts property (5) for $V_{(m+2)s+r}$ in the induction hypothesis.

Case 2. Assume $r \neq 0$, i.e., $r \in \{1, 2, \ldots, m + 1\}$.

Note that $V_{(m+2)s+r} \subseteq W^{m+1+r}(q_r)$, $f(q_{r+1}) = q_r$, and by Remark 2.7, $f(W^{m+1+r+1}(q_{r+1})) = W^{m+1+r}(q_r)$. We can choose, for each $x \in V_{(m+2)s+r}$, a point $x' \in W^{m+1+r+1}(q_{r+1})$ such that $f(x') = x$. Then define $V_{(m+2)s+r+1}$ to be the collection of all such chosen x' that corresponds to $x \in V_{(m+2)s+r}$. Properties (2), (3), and (4) are clear from the construction. To establish property (1) in the case when $r \in \{1, 2, \ldots, m - 1\}$, we recall that $v_{(m+2)s+r+1} = q_{r+1}$. For the case when $r \in \{m, m+1\}$, we note that $V_{(m+2)s+r+1} \subseteq W^{m+1+r+1}(q_{r+1})$ and $q_{r+1} \in W^r(f(p))$, so

$$V_{(m+2)s+r+1} \subseteq W^{2m}(q_{r+1}) \subseteq W^m(f(p)) = W^m(v_{(m+2)s+r+1}).$$

We only need to verify property (5) now. Indeed, suppose that distinct points $x, y \in V_{(m+2)s+r+1}$ satisfy $\{x, y\} \subseteq A$ for some $A \in \bigvee_{i=0}^{(m+2)s+r} f^{-i}(\mathbf{W}^l)$. Then by construction $f(x)$, $f(y)$, and $f(A)$ satisfy

(a) $f(A) \subseteq B$ for some $B \in \bigvee_{i=0}^{(m+2)s+r-1} f^{-i}(\mathbf{W}^l)$,
(b) $f(x), f(y) \in V_{(m+2)s+r}$, and $f(x) \neq f(y)$,
(c) $\{f(x), f(y)\} \subseteq f(A) \subseteq B$.

This contradicts property (5) for $V_{(m+2)s+r}$ in the induction hypothesis.

The recursive construction and the inductive proof of the properties of the construction are now complete.

For each $s \in \mathbb{N}$, we consider

$$B_{(m+2)s} = \bigcap_{j=0}^{(m+2)s-1} f^{-j}\big(W^m\big(v_{(m+2)s-j}\big)\big) \in \bigvee_{j=0}^{(m+2)s-1} f^{-j}(\mathbf{W}^m).$$

Then $V_{(m+2)s} \subseteq B_{(m+2)s}$ by properties (1) and (2) of the construction. On the other hand, by property (5), if $\mathscr{A} \subseteq \bigvee_{j=0}^{(m+2)s-1} f^{-j}(\mathbf{W}^l)$ satisfies

$$\bigcup \mathscr{A} \supseteq B_{(m+2)s} \supseteq V_{(m+2)s}.$$

So $\operatorname{card} \mathscr{A} \geq \operatorname{card} V_{(m+2)s} = k^s$, where the equality follows from property (4).

Thus by (3.14),

$$\begin{aligned} h(f|\mathbf{W}^m) &= \lim_{l\to+\infty} \lim_{n\to+\infty} \frac{1}{n} H\left(\bigvee_{i=0}^{n-1} f^{-i}(\mathbf{W}^l) \middle| \bigvee_{j=0}^{n-1} f^{-j}(\mathbf{W}^m)\right) \\ &\geq \liminf_{l\to+\infty} \liminf_{s\to+\infty} \frac{1}{(m+2)s} \log\left(k^s\right) = \frac{\log k}{m+2} > 0. \end{aligned}$$

Therefore, the map f is not h-expansive. □

Proof (Proof of Theorem 6.5*)* By Alaoglu's theorem, the space $\mathscr{M}(S^2, f)$ of f-invariant Borel probability measures equipped with the weak* topology is compact. Since the measure-theoretic entropy $\mu \mapsto h_\mu(f)$ is upper semi-continuous by Corollary 6.4, so is $\mu \mapsto P_\mu(f, \psi)$ by (3.4). Thus $\mu \mapsto P_\mu(f, \psi)$ attains its supremum over $\mathscr{M}(S^2, f)$ at a measure μ_ψ, which by the Variational Principle (3.5) is an equilibrium state for the map f and the potential ψ. □

Chapter 7
Large Deviation Principles

This chapter is devoted to the study of large deviation principles and equidistribution results for periodic points and iterated preimages of expanding Thurston maps without periodic critical points. The idea is to apply a general framework devised by Y. Kifer [Ki90] to obtain level-2 large deviation principles, and to derive the equidistribution results as consequences. The main theorem is the following.

Theorem 7.1 *Let $f\colon S^2 \to S^2$ be an expanding Thurston map with no periodic critical points, and d a visual metric on S^2 for f. Let $\mathcal{P}(S^2)$ denote the space of Borel probability measures on S^2 equipped with the weak* topology. Let ϕ be a real-valued Hölder continuous function on (S^2, d), and μ_ϕ be the unique equilibrium state for the map f and the potential ϕ.*

For each $n \in \mathbb{N}$, let $W_n\colon S^2 \to \mathcal{P}(S^2)$ be the continuous function defined by

$$W_n(x) = \frac{1}{n}\sum_{i=0}^{n-1} \delta_{f^i(x)},$$

and denote $S_n\phi(x) = \sum\limits_{i=0}^{n-1} \phi(f^i(x))$ for $x \in S^2$. Fix an arbitrary sequence of functions $\{w_n\colon S^2 \to \mathbb{R}\}_{n\in\mathbb{N}}$ satisfying $w_n(x) \in [1, \deg_{f^n}(x)]$ for each $n \in \mathbb{N}$ and each $x \in S^2$. We consider the following sequences of Borel probability measures on $\mathcal{P}(S^2)$:

***Iterated preimages**: Given a sequence $\{x_n\}_{n\in\mathbb{N}}$ of points in S^2, for each $n \in \mathbb{N}$, put*

$$\Omega_n(x_n) = \sum_{y\in f^{-n}(x_n)} \frac{w_n(y)\exp(S_n\phi(y))}{\sum_{z\in f^{-n}(x_n)} w_n(z)\exp(S_n\phi(z))}\delta_{W_n(y)}.$$

***Periodic points**: For each $n \in \mathbb{N}$, put*

$$\Omega_n = \sum_{x=f^n(x)} \frac{w_n(x)\exp(S_n\phi(x))}{\sum_{y=f^n(y)} w_n(y)\exp(S_n\phi(y))}\delta_{W_n(x)}.$$

Z. Li, *Ergodic Theory of Expanding Thurston Maps*, Atlantis Studies in Dynamical Systems 4, DOI 10.2991/978-94-6239-174-1_7

Then each of the sequences $\{\Omega_n(x_n)\}_{n\in\mathbb{N}}$ *and* $\{\Omega_n\}_{n\in\mathbb{N}}$ *converges to* δ_{μ_ϕ} *in the weak* topology, and satisfies a large deviation principle with rate function* $I^\phi : \mathscr{P}(S^2) \to [0, +\infty]$ *given by*

$$I^\phi(\mu) = \begin{cases} P(f,\phi) - \int \phi \, \mathrm{d}\mu - h_\mu(f) & \text{if } \mu \in \mathcal{M}(S^2, f); \\ +\infty & \text{if } \mu \in \mathscr{P}(S^2)\backslash \mathcal{M}(S^2, f). \end{cases} \tag{7.1}$$

Furthermore, for each convex open subset $\mathfrak{G}$ *of* $\mathscr{P}(S^2)$ *containing some invariant measure, we have*

$$-\inf_{\mathfrak{G}} I^\phi = \lim_{n\to+\infty} \frac{1}{n} \log \Omega_n(x_n)(\mathfrak{G}) = \lim_{n\to+\infty} \frac{1}{n} \log \Omega_n(\mathfrak{G}) \tag{7.2}$$

and (7.2) remains true with $\mathfrak{G}$ *replaced by its closure* $\overline{\mathfrak{G}}$*.*

See Sect. 7.1 for a brief introduction to large deviation principles in our context. As an immediate consequence, we get the following corollary. See Sect. 7.4 for the proof.

Corollary 7.2 *Let* $f: S^2 \to S^2$ *be an expanding Thurston map with no periodic critical points, and* d *a visual metric on* S^2 *for* f*. Let* ϕ *be a real-valued Hölder continuous function on* (S^2, d)*, and* μ_ϕ *be the unique equilibrium state for the map* f *and the potential* ϕ*. Given a sequence* $\{x_n\}_{n\in\mathbb{N}}$ *of points in* S^2*. Fix an arbitrary sequence of functions* $\{w_n : S^2 \to \mathbb{R}\}_{n\in\mathbb{N}}$ *satisfying* $w_n(x) \in [1, \deg_{f^n}(x)]$ *for each* $n \in \mathbb{N}$ *and each* $x \in S^2$*.*

Then for each $\mu \in \mathcal{M}(S^2, f)$*, and each convex local basis* G_μ *of* $\mathscr{P}(S^2)$ *at* μ*, we have*

$$\begin{aligned} h_\mu(f) + \int \phi \, \mathrm{d}\mu &= \inf\left\{ \lim_{n\to+\infty} \frac{1}{n} \log \sum_{y\in f^{-n}(x_n),\, W_n(y)\in\mathfrak{G}} w_n(y) e^{S_n\phi(y)} \,\middle|\, \mathfrak{G} \in G_\mu \right\} \\ &= \inf\left\{ \lim_{n\to+\infty} \frac{1}{n} \log \sum_{x=f^n(x),\, W_n(x)\in\mathfrak{G}} w_n(x) e^{S_n\phi(x)} \,\middle|\, \mathfrak{G} \in G_\mu \right\}. \end{aligned} \tag{7.3}$$

Here W_n *and* $S_n\phi$ *are as defined in Theorem 7.1.*

Equidistribution results follow from corresponding level-2 large deviation principles.

Corollary 7.3 *Let* $f: S^2 \to S^2$ *be an expanding Thurston map with no periodic critical points, and* d *a visual metric on* S^2 *for* f*. Let* ϕ *be a real-valued Hölder continuous function on* (S^2, d)*, and* μ_ϕ *be the unique equilibrium state for the map* f *and the potential* ϕ*. Fix an arbitrary sequence of functions* $\{w_n : S^2 \to \mathbb{R}\}_{n\in\mathbb{N}}$ *satisfying* $w_n(x) \in [1, \deg_{f^n}(x)]$ *for each* $n \in \mathbb{N}$ *and each* $x \in S^2$*.*

We consider the following sequences of Borel probability measures on S^2*:*

***Iterated preimages**: Given a sequence $\{x_n\}_{n\in\mathbb{N}}$ of points in S^2, for each $n \in \mathbb{N}$, put*

$$\nu_n = \sum_{y\in f^{-n}(x_n)} \frac{w_n(y)\exp(S_n\phi(y))}{\sum_{z\in f^{-n}(x_n)} w_n(z)\exp(S_n\phi(z))} \frac{1}{n}\sum_{i=0}^{n-1}\delta_{f^i(y)},$$

***Periodic points**: For each $n \in \mathbb{N}$, put*

$$\eta_n = \sum_{x=f^n(x)} \frac{w_n(x)\exp(S_n\phi(x))}{\sum_{y=f^n(y)} w_n(y)\exp(S_n\phi(y))} \frac{1}{n}\sum_{i=0}^{n-1}\delta_{f^i(x)}.$$

Then as $n \longrightarrow +\infty$,

$$\nu_n \xrightarrow{w^*} \mu_\phi \quad and \quad \eta_n \xrightarrow{w^*} \mu_\phi.$$

Here S_n is defined as in Theorem 7.1.

Remark 7.4 Since $S_n\phi(f^i(x)) = S_n\phi(x)$ for $i \in \mathbb{N}$ if $f^n(x) = x$, we get

$$\eta_n = \sum_{x=f^n(x)} \frac{\frac{S_n w_n(x)}{n}\exp(S_n\phi(x))}{\sum_{y=f^n(y)} w_n(y)\exp(S_n\phi(y))}\delta_x,$$

for $n \in \mathbb{N}$. In particular, when $w_n(\cdot) \equiv 1$,

$$\eta_n = \sum_{x=f^n(x)} \frac{\exp(S_n\phi(x))}{\sum_{y=f^n(y)} \exp(S_n\phi(y))}\delta_x;$$

when $w_n(x) = \deg_{f^n}(x)$, since $\deg_{f^n}(f^i(x)) = \deg_{f^n}(x)$ for $i \in \mathbb{N}$ if $f^n(x) = x$, we have

$$\eta_n = \sum_{x=f^n(x)} \frac{\deg_{f^n}(x)\exp(S_n\phi(x))}{\sum_{y=f^n(y)} \deg_{f^n}(y)\exp(S_n\phi(y))}\delta_x.$$

See Sect. 7.4 for the proof of Corollary 7.3. Note that the part of Corollary 7.3 on iterated preimages generalizes (5.102) and (5.103) in Proposition 5.54 in the context of expanding Thurston maps without periodic critical points. We also remark that our results Corollary 6.4 through Corollary 7.3 are only known in this context. In particular, the following questions for expanding Thurston maps $f\colon S^2 \to S^2$ with at least one periodic critical point are still open.

Question 7.5 Is the measure-theoretic entropy $\mu \mapsto h_\mu(f)$ upper semi-continuous?

Question 7.6 Are iterated preimages and periodic points equidistributed with respect to the unique equilibrium state for a Hölder continuous potential?

Note that regarding Question 2, we know that iterated preimages, counted with local degree, are equidistributed with respect to the equilibrium state by (5.102) in Proposition 5.54. If Question 1 can be answered positively, then the mechanism of Theorem 7.7 works and we get that the equidistribution of periodic points from the corresponding large deviation principle. However, for iterated preimages without counting local degree (i.e., when $w_n(\cdot) \neq \deg_{f^n}(\cdot)$ in Corollary 7.3, and in particular, when $w_n(\cdot) \equiv 1$), the verification of Condition (2) mentioned earlier for Theorem 7.7 to apply still remains unknown. Compare (7.6) and (7.7) in Proposition 7.8.

In Sect. 7.1, we give a brief review of level-2 large deviation principles in our context. We record the theorem of Y. Kifer [Ki90], reformulated by H. Comman and J. Rivera-Letelier [CRL11], on level-2 large deviation principles. This result, stated in Theorem 7.7, will be applied later to our context.

We generalize some characterization of topological pressure in Sect. 7.2 in our context. More precisely, we use equidistribution results for iterated preimages in Proposition 5.54 to show in Propositions 7.8 and 7.9 that

$$P(f,\phi) = \lim_{n\to+\infty} \frac{1}{n} \log \sum w_n(y) \exp(S_n\phi(y)), \tag{7.4}$$

where the sum is taken over preimages under f^n in Proposition 7.8, and over periodic points in Proposition 7.9, the potential $\phi\colon S^2 \to \mathbb{R}$ is Hölder continuous with respect to a visual metric d, and the weight $w_n(y) \in [1, \deg_{f^n}(y)]$ for $n \in \mathbb{N}$ and $y \in S^2$. We note that for periodic points, the equation (7.4) is established in Proposition 7.9 for all expanding Thurston maps, but for iterated preimages, we only obtain (7.4) for expanding Thurston maps without periodic critical points in Proposition 7.8.

In Sect. 7.3, by applying Theorem 7.7 to give a proof of Theorem 7.1, we finally establish level-2 large deviation principles in the context of expanding Thurston maps without periodic critical points and given Hölder continuous potentials.

Section 7.4 consists of the proofs of Corollaries 7.2 and 7.3. We first obtain characterizations of the measure-theoretic pressure in terms of the infimum of certain limits involving periodic points and iterated preimages (Corollary 7.2). Such characterizations are then used in the proof of the equidistribution results (Corollary 7.3).

7.1 Level-2 Large Deviation Principles

Let X be a compact metrizable topological space. Recall that $\mathcal{P}(X)$ is the set of Borel probability measures on X. We equip $\mathcal{P}(X)$ with the weak* topology. Note that this topology is metrizable (see for example, [Con85, Theorem 5.1]). Let I : $\mathcal{P}(X) \to [0, +\infty]$ be a *lower semi-continuous function*, i.e., I satisfy the condition that $\liminf_{y\to x} I(y) \geq I(x)$ for all $x \in \mathcal{P}(X)$.

A sequence $\{\Omega_n\}_{n\in\mathbb{N}}$ of Borel probability measures on $\mathcal{P}(X)$ is said to satisfy a *large deviation principle with rate function I* if for each closed subset $\mathfrak{F}$ of $\mathcal{P}(X)$

and each open subset $\mathfrak{G}$ of $\mathscr{P}(X)$ we have

$$\limsup_{n\to+\infty} \frac{1}{n} \log \Omega_n(\mathfrak{F}) \leq -\inf\{I(x) \mid x \in \mathfrak{F}\},$$

and

$$\liminf_{n\to+\infty} \frac{1}{n} \log \Omega_n(\mathfrak{G}) \geq -\inf\{I(x) \mid x \in \mathfrak{G}\}.$$

We will apply the following theorem due to Y. Kifer [Ki90, Theorem 4.3], reformulated by H. Comman and J. Rivera-Letelier [CRL11, Theorem C].

Theorem 7.7 (Y. Kifer; H. Comman and J. Rivera-Letelier) *Let X be a compact metrizable topological space, and let $g\colon X \to X$ be a continuous map. Fix $\phi \in C(X)$, and let H be a dense vector subspace of $C(X)$ with respect to the uniform norm. Let $I^\phi\colon \mathscr{P}(X) \to [0, +\infty]$ be the function defined by*

$$I^\phi(\mu) = \begin{cases} P(g,\phi) - \int \phi \, \mathrm{d}\mu - h_\mu(g) & \text{if } \mu \in \mathscr{M}(X,g); \\ +\infty & \text{if } \mu \in \mathscr{P}(X) \setminus \mathscr{M}(X,g). \end{cases}$$

We assume the following conditions are satisfied:

(i) The measure-theoretic entropy $h_\mu(g)$ of g, as a function of μ defined on $\mathscr{M}(X,g)$ (equipped with the weak topology), is finite and upper semi-continuous.*
(ii) For each $\psi \in H$, there exists a unique equilibrium state for the map g and the potential $\phi + \psi$.

Then every sequence $\{\Omega_n\}_{n\in\mathbb{N}}$ of Borel probability measures on $\mathscr{P}(X)$ such that for each $\psi \in H$,

$$\lim_{n\to+\infty} \frac{1}{n} \log \int_{\mathscr{P}(X)} \exp\left(n \int \psi \, \mathrm{d}\mu\right) \mathrm{d}\Omega_n(\mu) = P(g, \phi + \psi) - P(g, \phi), \tag{7.5}$$

satisfies a large deviation principle with rate function I^ϕ, and it converges in the weak topology to the Dirac measure supported on the unique equilibrium state for the map g and the potential ϕ. Furthermore, for each convex open subset $\mathfrak{G}$ of $\mathscr{P}(X)$ containing some invariant measure, we have*

$$\lim_{n\to+\infty} \frac{1}{n} \log \Omega_n(\mathfrak{G}) = \lim_{n\to+\infty} \frac{1}{n} \log \Omega_n\left(\overline{\mathfrak{G}}\right) = -\inf_{\mathfrak{G}} I^\phi = -\inf_{\overline{\mathfrak{G}}} I^\phi.$$

Recall that $P(g,\phi)$ is the topological pressure of the map g with respect to the potential ϕ.

In our context, $X = S^2$, the map $g = f$ where $f\colon S^2 \to S^2$ is an expanding Thurston map with no periodic critical points. Fix a visual metric d on S^2 for f. The function ϕ is a real-valued Hölder continuous function with an exponent $\alpha \in (0,1]$. Then $H = C^{0,\alpha}(S^2, d)$ is the space of real-valued Hölder continuous functions with the exponent α on (S^2, d). Note that $C^{0,\alpha}(S^2, d)$ is dense in $C(S^2)$ (equipped with the uniform norm) (Lemma 5.34). Condition (i) is satisfied by Corollary 6.4. Condition (ii) is guaranteed by Theorem 5.1. Thus we just need to verify (7.5) for the sequences that we will consider in this chapter.

7.2 Characterizations of the Pressure $P(f, \phi)$

Let f, d, ϕ, α satisfy the Assumptions. Recall that m_ϕ is the unique eigenmeasure of $\mathcal{L}_\phi^*$, i.e., the unique Borel probability measure on S^2 that satisfies $\mathcal{L}_\phi^*(m_\phi) = cm_\phi$ for some constant $c \in \mathbb{R}$ (compare Theorem 5.12 and Corollary 5.32).

We now prove a slight generalization of Proposition 5.19.

Proposition 7.8 *Let f, d, ϕ, α satisfy the Assumptions. Then for each sequence $\{x_n\}_{n\in\mathbb{N}}$ in S^2, we have*

$$P(f, \phi) = \lim_{n\to+\infty} \frac{1}{n} \log \sum_{y\in f^{-n}(x_n)} \deg_{f^n}(y) \exp(S_n\phi(y)). \tag{7.6}$$

If we also assume that f has no periodic critical points, then for an arbitrary sequence of functions $\{w_n : S^2 \to \mathbb{R}\}_{n\in\mathbb{N}}$ satisfying $w_n(x) \in [1, \deg_{f^n}(x)]$ for each $n \in \mathbb{N}$ and each $x \in S^2$, we have

$$P(f, \phi) = \lim_{n\to+\infty} \frac{1}{n} \log \sum_{y\in f^{-n}(x_n)} w_n(y) \exp(S_n\phi(y)). \tag{7.7}$$

Proof We fix a Jordan curve $\mathcal{C} \subseteq S^2$ that satisfies the Assumptions (see Theorem 2.16 for the existence of such $\mathcal{C}$). By Proposition 5.9, for each $x \in S^2$ we have

$$P(f, \phi) = \lim_{n\to+\infty} \frac{1}{n} \log \sum_{y\in f^{-n}(x)} \deg_{f^n}(y) \exp(S_n\phi(y)).$$

Combining this equation with (5.6) in Lemma 5.4, we get (7.6).

Assume now that f has no periodic critical points. Then there exists a finite number $M \in \mathbb{N}$ that depends only on f such that $\deg_{f^n}(x) \leq M$ for $n \in \mathbb{N}_0$ and $x \in S^2$ [BM17, Lemma 18.6]. Thus for each $n \in \mathbb{N}$,

$$1 \leq \frac{\sum_{y\in f^{-n}(x_n)} \deg_{f^n}(y) \exp(S_n\phi(y))}{\sum_{y\in f^{-n}(x_n)} w_n(y) \exp(S_n\phi(y))} \leq M.$$

Hence (7.7) follows from (7.6). □

While Proposition 7.8 is a statement for iterated preimages, the next proposition is for periodic points. Recall that $P_{1,f^n} = \{x \in S^2 \mid f^n(x) = x\}$ for $n \in \mathbb{N}$.

Proposition 7.9 *Let f, d, ϕ, α satisfy the Assumptions. Fix an arbitrary sequence of functions $\{w_n : S^2 \to \mathbb{R}\}_{n\in\mathbb{N}}$ satisfying $w_n(x) \in [1, \deg_{f^n}(x)]$ for each $n \in \mathbb{N}$ and each $x \in S^2$. Then*

$$P(f, \phi) = \lim_{n\to+\infty} \frac{1}{n} \log \sum_{x\in P_{1,f^n}} w_n(x) \exp(S_n\phi(x)). \tag{7.8}$$

Proof We fix a Jordan curve $\mathscr{C} \subseteq S^2$ that satisfies the Assumptions (see Theorem 2.16 for the existence of such $\mathscr{C}$).

By Proposition 7.8, it suffices to prove that there exist $C > 1$ and $z \in S^2$ such that for each $n \in \mathbb{N}$ sufficiently large,

$$\frac{1}{C} \leq \frac{\sum\limits_{x\in P_{1,f^n}} w_n(x) \exp(S_n\phi(x))}{\sum\limits_{x\in f^{-n}(z)} \deg_{f^n}(x) \exp(S_n\phi(x))} \leq C. \tag{7.9}$$

We fix a 0-edge $e_0 \subseteq \mathscr{C}$ and a point $z \in \operatorname{inte}(e_0)$.

By Proposition 5.39, $m_\phi(\mathscr{C}) = 0$. By the continuity of m_ϕ, we can find $\delta > 0$ such that

$$m_\phi\big(\overline{N_d^\delta(\mathscr{C})}\big) < \frac{1}{100}. \tag{7.10}$$

Note that $\deg_{f^n}(y) = 1$ if $f^n(y) = z$ for $n \in \mathbb{N}$. We define, for each $n \in \mathbb{N}_0$, the probability measure

$$\nu_n = \sum_{x\in f^{-n}(z)} \frac{\deg_{f^n}(x) \exp(S_n\phi(x))}{\sum_{y\in f^{-n}(z)} \deg_{f^n}(y) \exp(S_n\phi(y))} \delta_x = \sum_{x\in f^{-n}(z)} \frac{\exp(S_n\phi(x))}{\sum_{y\in f^{-n}(z)} \exp(S_n\phi(y))} \delta_x. \tag{7.11}$$

Let $N_0 \in \mathbb{N}$ be the constant from Lemma 4.9. By (5.101) in Proposition 5.54, $\nu_n \xrightarrow{w^*} m_\phi$ as $n \longrightarrow +\infty$. So by Lemma 4.19, we can choose $N_1 > N_0$ such that for each $n \in \mathbb{N}$ with $n > N_1$, we have

$$\nu_n\big(\overline{N_d^\delta(\mathscr{C})}\big) < \frac{1}{10}. \tag{7.12}$$

By Lemma 2.13, it is clear that we can choose $N_2 > N_1$ such that for each $n \in \mathbb{N}$ with $n > N_2$, and each n-tile $X^n \in \mathbf{X}^n$,

$$\operatorname{diam}_d(X^n) < \frac{\delta}{10}. \tag{7.13}$$

We observe that for each $i \in \mathbb{N}$, we can pair a white i-tile $X^i_w \in \mathbf{X}^i_w$ and a black i-tile $X^i_b \in \mathbf{X}^i_b$ whose intersection $X^i_w \cap X^i_b$ is an i-edge contained in $f^{-i}(e_0)$. There are a total of $(\deg f)^i$ such pairs and each i-tile is in exactly one such pair. We denote by $\mathbf{P}_i$ the collection of the unions $X^i_w \cup X^i_b$ of such pairs, i.e.,

$$\mathbf{P}_i = \{X^i_w \cup X^i_b \mid X^i_w \in \mathbf{X}^i_w, X^i_b \in \mathbf{X}^i_b, X^i_w \cap X^i_b \cap f^{-i}(e_0) \in \mathbf{E}^i\}.$$

We denote $\mathbf{P}^\delta_i = \{A \in \mathbf{P}_i \mid A \setminus N^\delta_d(\mathscr{C}) \neq \emptyset\}$.

We now fix an integer $n > N_2$.

Then $\mathbf{P}^\delta_n$ forms a cover of $S^2 \setminus \overline{N^\delta_d(\mathscr{C})}$. For each $A \in \mathbf{P}^\delta_n$, by (7.13) we have $A \cap \mathscr{C} = \emptyset$. So $A \subseteq \operatorname{inte} X^0_w$ or $A \subseteq \operatorname{inte} X^0_b$, where X^0_w and X^0_b are the white 0-tile and the black 0-tile in $\mathbf{X}^0$, respectively. So by Brouwer's Fixed Point Theorem (see for example, [Ha02, Theorem 1.9]) and Lemma 4.9, we can define a function $p\colon \mathbf{P}^\delta_n \to P_{1,f^n}$ in such a way that $p(A)$ is the unique fixed point of f^n contained in A. (For example, if $A \in \mathbf{P}^\delta_n$ is the union of a black n-tile X^n_b and a white n-tile X^n_w and is a subset of the interior of the black 0-tile, then there is no fixed point of f^n in X^n_w, and by applying Brouwer's Fixed Point Theorem to the inverse of f^n restricted to X^n_b, we get a fixed point $x \in X^n_b$ of f^n, which is the unique fixed point of f^n in X^n_b by Lemma 4.9.) Moreover, for each $A \in \mathbf{P}^\delta_n$, $p(A) \in \operatorname{int} A$, so $\deg_{f^n}(p(A)) = 1 = w_n(p(A))$. In general, by Lemma 4.9, each $A \in \mathbf{P}_n$ contains at most 2 fixed points of f^n.

We also define a function $q\colon \mathbf{P}_n \to f^{-n}(z)$ in such a way that $q(A)$ is the unique preimage of z under f^n that is contained in A, for each $A \in \mathbf{P}_n$ (see Proposition 2.6). We note that if $X^n_w \in \mathbf{X}^n_w$ and $X^n_b \in \mathbf{X}^n_b$ are the n-tiles that satisfy $X^n_w \cup X^n_b = A \in \mathbf{P}_n$ and $e_n = X^n_w \cap X^n_b$, then $q(A) \in e_n$. Thus in particular, $\deg_{f^n}(q(A)) = 1$ for each $A \in \mathbf{P}_n$.

Hence by construction, we have

$$\sum_{x \in f^{-n}(z)} e^{S_n\phi(x)} = \sum_{A \in \mathbf{P}^\delta_n} e^{S_n\phi(q(A))} + \sum_{A \in \mathbf{P}_n \setminus \mathbf{P}^\delta_n} e^{S_n\phi(q(A))}, \tag{7.14}$$

and

$$\sum_{A \in \mathbf{P}^\delta_n} e^{S_n\phi(p(A))} \le \sum_{x \in P_{1,f^n}} w_n(x) e^{S_n\phi(x)} \le \sum_{A \in \mathbf{P}^\delta_n} e^{S_n\phi(p(A))} + \sum_{A \in \mathbf{P}_n \setminus \mathbf{P}^\delta_n} \sum_{x \in A \cap P_{1,f^n}} e^{S_n\phi(x)}. \tag{7.15}$$

The last inequality in (7.15) is due to the fact that if $x \in P_{1,f^n}$ satisfies $\deg_{f^n}(x) \ge 2$, then $x \in \mathbf{V}^n$ with $x \notin \bigcup \mathbf{P}^\delta_n$, and the number of $A \in \mathbf{P}_n$ that contains x is at least $\deg_{f^n}(x)$ (and at most $2\deg_{f^n}(x)$).

By (5.4) in Lemma 5.3, we get

$$\frac{1}{C_3} \le \frac{\sum_{A \in \mathbf{P}^\delta_n} e^{S_n\phi(p(A))}}{\sum_{A \in \mathbf{P}^\delta_n} e^{S_n\phi(q(A))}} \le C_3, \tag{7.16}$$

and since in addition, $\mathrm{card}(A \cap P_{1,f^n}) \leq 2$ for $A \in \mathbf{P}_n$ by Lemma 4.9, we have

$$\frac{\sum\limits_{A\in \mathbf{P}_n \backslash \mathbf{P}_n^\delta} \sum\limits_{x\in A\cap P_{1,f^n}} e^{S_n\phi(x)}}{\sum\limits_{A\in \mathbf{P}_n \backslash \mathbf{P}_n^\delta} e^{S_n\phi(q(A))}} \leq 2C_3, \tag{7.17}$$

where

$$C_3 = \exp\left(C_1 \left(\mathrm{diam}_d(S^2)\right)^\alpha\right),$$

and $C_1 > 0$ is a constant from Lemma 5.3. Both C_1 and C_3 depend only on f, $\mathcal{C}$, d, ϕ, and α.

By (7.14), (7.11), and (7.12), we get

$$\sum_{x\in f^{-n}(z)} e^{S_n\phi(x)} \geq \sum_{A\in \mathbf{P}_n^\delta} e^{S_n\phi(q(A))} \geq \frac{9}{10} \sum_{x\in f^{-n}(z)} e^{S_n\phi(x)}. \tag{7.18}$$

Hence, by (7.15), (7.16), and (7.18), we have

$$\frac{\sum\limits_{x\in P_{1,f^n}} w_n(x) e^{S_n\phi(x)}}{\sum\limits_{x\in f^{-n}(z)} \deg_{f^n}(x) e^{S_n\phi(x)}} \geq \frac{\sum\limits_{A\in \mathbf{P}_n^\delta} e^{S_n\phi(p(A))}}{\frac{10}{9}\sum\limits_{A\in \mathbf{P}_n^\delta} e^{S_n\phi(q(A))}} \geq \frac{9}{10C_3}.$$

On the other hand, by (7.14), (7.15), (7.16), (7.17), and (7.18), we get

$$\begin{aligned}
\frac{\sum\limits_{x\in P_{1,f^n}} w_n(x) e^{S_n\phi(x)}}{\sum\limits_{x\in f^{-n}(z)} \deg_{f^n}(x) e^{S_n\phi(x)}} &\leq \frac{\sum\limits_{A\in \mathbf{P}_n^\delta} e^{S_n\phi(p(A))} + \sum\limits_{A\in \mathbf{P}_n \backslash \mathbf{P}_n^\delta} \sum\limits_{x\in A\cap P_{1,f^n}} e^{S_n\phi(x)}}{\sum\limits_{x\in f^{-n}(z)} e^{S_n\phi(x)}} \\
&\leq \frac{\sum\limits_{A\in \mathbf{P}_n^\delta} e^{S_n\phi(p(A))}}{\sum\limits_{A\in \mathbf{P}_n^\delta} e^{S_n\phi(q(A))}} + \frac{\sum\limits_{A\in \mathbf{P}_n \backslash \mathbf{P}_n^\delta} \sum\limits_{x\in A\cap P_{1,f^n}} e^{S_n\phi(x)}}{10 \sum\limits_{A\in \mathbf{P}_n \backslash \mathbf{P}_n^\delta} e^{S_n\phi(q(A))}} \leq C_3 + \frac{2}{10} C_3.
\end{aligned}$$

Thus (7.9) holds if we choose $C = 2C_3$ and $n > N_2$. The proof is now complete. ☐

7.3 Proof of Large Deviation Principles

Proof (*Proof of Theorem* 7.1) Let $\phi \in C^{0,\alpha}(S^2, d)$ for some $\alpha \in (0, 1]$.

We apply Theorem 7.7 with $X = S^2$, $g = f$, and $H = C^{0,\alpha}(S^2, d)$. Note that $C^{0,\alpha}(S^2, d)$ is dense in $C(S^2)$ with respect to the uniform norm (Lemma 5.34).

Theorem 5.1 implies Condition (ii) in the hypothesis of Theorem 7.7. Condition (i) follows from Corollary 6.4, (3.6), and the fact that $h_{\mathrm{top}}(f) = \log(\deg f)$ ([BM17], Corollary 17.2).

It now suffices to verify (7.5) for each of the sequences $\{\Omega_n(x_n)\}_{n\in\mathbb{N}}$ and $\{\Omega_n\}_{n\in\mathbb{N}}$ of Borel probability measures on $\mathcal{P}(S^2)$.

Fix an arbitrary $\psi \in C^{0,\alpha}(S^2, d)$.

By (7.7) in Proposition 7.8,

$$\begin{aligned}
&\lim_{n\to+\infty} \frac{1}{n} \log \int_{\mathcal{P}(S^2)} \exp\left(n \int \psi \,\mathrm{d}\mu\right) \mathrm{d}\Omega_n(x_n)(\mu) \\
=& \lim_{n\to+\infty} \frac{1}{n} \log \sum_{y\in f^{-n}(x_n)} \frac{w_n(y)\exp(S_n\phi(y))}{\sum_{z\in f^{-n}(x_n)} w_n(z)\exp(S_n\phi(z))} e^{\sum_{i=0}^{n-1}\psi(f^i(y))} \\
=& \lim_{n\to+\infty} \frac{1}{n} \left(\log \sum_{y\in f^{-n}(x_n)} w_n(y) e^{S_n(\phi+\psi)(y)} - \log \sum_{z\in f^{-n}(x_n)} w_n(z) e^{S_n(\phi)(z)} \right) \\
=& P(f, \phi+\psi) - P(f,\phi).
\end{aligned}$$

Similarly, by (7.8) in Proposition 7.9, we get

$$P(f,\phi+\psi) - P(f,\phi) = \lim_{n\to+\infty} \frac{1}{n} \log \int_{\mathcal{P}(S^2)} \exp\left(n \int \psi \,\mathrm{d}\mu\right) \mathrm{d}\Omega_n(\mu)$$

The theorem now follows from Theorem 7.7. □

7.4 Equidistribution Revisited

Proof (*Proof of Corollary* 7.2) We prove the first equality in (7.3) now.

Fix $\mu \in \mathcal{M}(S^2, f)$ and a convex local basis G_μ at μ. By (7.1) and the upper semi-continuity of $h_\mu(f)$ (Corollary 6.4), we get

$$-I^\phi(\mu) = \inf_{\mathfrak{G}\in G_\mu} \left(\sup_{\mathfrak{G}} (-I^\phi) \right) = \inf_{\mathfrak{G}\in G_\mu} \left(-\inf_{\mathfrak{G}} I^\phi \right).$$

Then by (7.1) and (7.2),

$$\begin{aligned}
&- P(f,\phi) + \int \phi \,\mathrm{d}\mu + h_\mu(f) = -I^\phi(\mu) = \inf_{\mathfrak{G}\in G_\mu} \left(-\inf_{\mathfrak{G}} I^\phi \right) \\
=& \inf_{\mathfrak{G}\in G_\mu} \left\{ \lim_{n\to+\infty} \frac{1}{n} \log \sum_{y\in f^{-n}(x_n),\, W_n(y)\in\mathfrak{G}} \frac{w_n(y)\exp(S_n\phi(y))}{Z_n(\phi)} \right\},
\end{aligned}$$

where we write $Z_n(\phi) = \sum_{z\in f^{-n}(x_n)} w_n(z)\exp(S_n\phi(z))$. By (7.7) in Proposition 7.8, we have $P(f,\phi) = \lim_{n\to+\infty} \frac{1}{n}\log Z_n(\phi)$. Thus the first equality in (7.3) follows.

By similar arguments, with (7.6) in Proposition 7.8 replaced by (7.8) in Proposition 7.9, we get the second equality in (7.3). □

Proof (*Proof of Corollary* 7.3) Recall that $W_n(x) = \frac{1}{n}\sum_{i=0}^{n-1} \delta_{f^i(x)} \in \mathcal{P}(S^2)$ for $x \in S^2$ and $n \in \mathbb{N}$ as defined in (0.4). We write

$$Z_n^+(\mathfrak{G}) = \sum_{y\in f^{-n}(x_n),\ W_n(y)\in\mathfrak{G}} \deg_{f^n}(y)\exp(S_n\phi(y))$$

and

$$Z_n^-(\mathfrak{G}) = \sum_{y\in f^{-n}(x_n),\ W_n(y)\notin\mathfrak{G}} \deg_{f^n}(y)\exp(S_n\phi(y))$$

for each $n \in \mathbb{N}$ and each open set $\mathfrak{G} \subseteq \mathcal{P}(S^2)$.

Let G_{μ_ϕ} be a convex local basis of $\mathcal{P}(S^2)$ at μ_ϕ. Fix an arbitrary convex open set $\mathfrak{G} \in G_{\mu_\phi}$.

By the uniqueness of the equilibrium state in our context and Corollary 7.2, we get that for each $\mu \in \mathcal{P}(S^2) \setminus \{\mu_\phi\}$, there exist numbers $a_\mu < P(f,\phi)$ and $N_\mu \in \mathbb{N}$ and an open neighborhood $\mathfrak{U}_\mu \subseteq \mathcal{P}(S^2) \setminus \{\mu_\phi\}$ containing μ such that for each $n > N_\mu$,

$$Z_n^+(\mathfrak{U}_\mu) \leq \exp(na_\mu). \tag{7.19}$$

Since $\mathcal{P}(S^2)$ is compact in the weak* topology by Alaoglu's theorem, so is $\mathcal{P}(S^2)\backslash\mathfrak{G}$. Thus there exists a finite set $\{\mu_i \mid i \in I\} \subseteq \mathcal{P}(S^2)\backslash\mathfrak{G}$ such that

$$\mathcal{P}(S^2)\backslash\mathfrak{G} \subseteq \bigcup_{i\in I} \mathfrak{U}_{\mu_i}. \tag{7.20}$$

Here I is a finite index set. Let $a = \max\{a_{\mu_i} \mid i \in I\}$. Note that $a < P(f,\phi)$. By Corollary 7.2 with $\mu = \mu_\phi$, we get that

$$P(f,\phi) \leq \lim_{n\to+\infty} \frac{1}{n}\log Z_n^+(\mathfrak{G}). \tag{7.21}$$

Combining (7.21) with (7.6) in Proposition 7.8, we get that the equality holds in (7.21). So there exist numbers $b \in (a, P(f,\phi))$ and $N \geq \max\{N_i \mid i \in I\}$ such that for each $n > N$,

$$Z_n^+(\mathfrak{G}) \geq \exp(nb). \tag{7.22}$$

We claim that every subsequential limit of $\{\nu_n\}_{n\in\mathbb{N}}$ in the weak* topology lies in the closure $\overline{\mathfrak{G}}$ of $\mathfrak{G}$. Assuming that the claim holds, then since $\mathfrak{G} \in G_{\mu_\phi}$ is arbitrary, we get that any subsequential limit of $\{\nu_n\}_{n\in\mathbb{N}}$ in the weak* topology is μ_ϕ, i.e., $\nu_n \xrightarrow{w^*} \mu_\phi$ as $n \longrightarrow +\infty$.

We now prove the claim. We first observe that for each $n \in \mathbb{N}$,

$$\begin{aligned}\nu_n &= \sum_{y\in f^{-n}(x_n)} \frac{w_n(y)\exp(S_n\phi(y))}{Z_n^+(\mathfrak{G}) + Z_n^-(\mathfrak{G})} W_n(y) \\ &= \frac{Z_n^+(\mathfrak{G})}{Z_n^+(\mathfrak{G}) + Z_n^-(\mathfrak{G})}\nu_n' + \sum_{y\in f^{-n}(x_n),\, W_n(y)\notin\mathfrak{G}} \frac{w_n(y)e^{S_n\phi(y)}}{Z_n^+(\mathfrak{G}) + Z_n^-(\mathfrak{G})} W_n(y),\end{aligned}$$

where $\nu_n' = \sum\limits_{y\in f^{-n}(x_n),\, W_n(y)\in\mathfrak{G}} \frac{w_n(y)\exp(S_n\phi(y))}{Z_n^+(\mathfrak{G})} W_n(y)$.

Note that since $a < b$, by (7.20), (7.19), and (7.22),

$$0 \le \lim_{n\to+\infty} \frac{Z_n^-(\mathfrak{G})}{Z_n^+(\mathfrak{G})} \le \lim_{n\to+\infty} \frac{\sum_{i\in I} Z_n^+(\mathfrak{U}_i)}{Z_n^+(\mathfrak{G})} \le \lim_{n\to+\infty} \frac{\operatorname{card}(I)\exp(na)}{\exp(nb)} = 0.$$

So $\lim\limits_{n\to+\infty} \frac{Z_n^+(\mathfrak{G})}{Z_n^+(\mathfrak{G})+Z_n^-(\mathfrak{G})} = 1$, and that the total variation

$$\begin{aligned}&\left\| \sum_{y\in f^{-n}(x_n),\, W_n(y)\notin\mathfrak{G}} \frac{w_n(y)\exp(S_n\phi(y))}{Z_n^+(\mathfrak{G}) + Z_n^-(\mathfrak{G})} W_n(y)\right\| \\ \le &\frac{\sum\limits_{y\in f^{-n}(x_n),\, W_n(y)\notin\mathfrak{G}} w_n(y)\exp(S_n\phi(y))\,\|W_n(y)\|}{Z_n^+(\mathfrak{G}) + Z_n^-(\mathfrak{G})} \\ \le &\frac{Z_n^-(\mathfrak{G})}{Z_n^+(\mathfrak{G}) + Z_n^-(\mathfrak{G})} \longrightarrow 0\end{aligned}$$

as $n \longrightarrow +\infty$. Thus a measure is a subsequential limit of $\{\nu_n\}_{n\in\mathbb{N}}$ if and only if it is a subsequential limit of $\{\nu_n'\}_{n\in\mathbb{N}}$. Note that ν_n' is a convex combination of measures in $\mathfrak{G}$, and $\mathfrak{G}$ is convex, so $\nu_n' \in \mathfrak{G}$, for $n \in \mathbb{N}$. Hence each subsequential limit of $\{\nu_n\}_{n\in\mathbb{N}}$ lies in the closure $\overline{\mathfrak{G}}$ of $\mathfrak{G}$. The proof of the claim is complete now.

By similar arguments as in the proof of the convergence of $\{\nu_n\}_{n\in\mathbb{N}}$ above, with (7.7) in Proposition 7.8 replaced by (7.8) in Proposition 7.9, we get that $\eta_n \xrightarrow{w^*} \mu_\phi$ as $n \longrightarrow +\infty$. □

References

[Ah82] AHLFORS, L.V., *Collected papers*, Vol. I, Birkhäuser, Boston, 1982.

[Bak85] BAKER, A., *A concise introduction to the theory of numbers*, Cambridge Univ. Press, Cambridge, 1985.

[BD11] BAKER, M. and DEMARCO, L., Preperiodic points and unlikely intersections. *Duke Math. J.* 159 (2011), 1–29.

[BJG09] BANG-JENSEN, J. and GUTIN, G., *Digraphs: theory, algorithms and applications*, Springer, London, 2009.

[Barn88] BARNSLEY, M., *Fractals everywhere*, Academic Press Professional, San Diego, 1988.

[Barr11] BARREIRA, L., *Thermodynamic formalism and applications to dimension theory*, Birkhäuser, Basel, 2011.

[Bi95] BILLINGSLEY, P., *Probability and measure*, John Wiley & Sons, New York, 1995.

[Bi99] BILLINGSLEY, P., *Convergence of probability measures*, John Wiley & Sons, New York, 1999.

[Bon06] BONK, M., Quasiconformal geometry of fractals. *Proc. Internat. Congr. Math., (Madrid, Spain, 2006)*, Europ. Math. Soc., Zürich, 2006, pp. 1349–1373.

[Bon11] BONK, M., Uniformization of Sierpinski carpets in the plane. *Invent. Math.* 186 (2011), 559–665.

[BonK02] BONK, M. and KLEINER, B., Quasisymmetric parametrizations of two-dimensional metric spheres. *Invent. Math.* 150 (2002), 127–183.

[BonK05] BONK, M. and KLEINER, B., Conformal dimension and Gromov hyperbolic groups with 2-sphere boundary. *Geom. Topol.* 9 (2005), 216–246.

[BKM09] BONK, M., KLEINER, B., and MERENKOV, S., Rigidity of Schottky sets. *Amer. J. Math.* 131 (2009), 409–443.

[BM10] BONK, M. and MEYER, D., Expanding Thurston maps. Preprint, (arXiv:1009.3647v1), 2010.

[BM17] BONK, M. and MEYER, D., *Expanding Thurston maps*, volume of *Mathematical Surveys and Monographs*, Amer. Math. Soc., Providence, RI, 2017.

[BouK13] BOURDON, M. and KLEINER, B., Combinatorial modulus, the combinatorial Loewner property, and Coxeter groups, *Groups Geom. Dyn.* 7 (2013), 39–107.

[Bow72] BOWEN, R., Entropy-expansive maps. *Trans. Amer. Math. Soc.* 164 (1972), 323–331.

[Bow75] BOWEN, R., *Equilibrium states and the ergodic theory of Anosov diffeomorphisms*, volume 470 of *Lect. Notes in Math.*, Springer, Berlin, 1975.

[Br10] BRACCI, F., Local holomorphic dynamics of diffeomorphisms in dimension one. In CONTRERAS, M.D. and DÍAZ-MADRIGAL, S. (Eds.), *Five lectures in complex analysis* (pp. 1–42), Amer. Math. Soc., Providence, RI, 2010.

Z. Li, *Ergodic Theory of Expanding Thurston Maps*, Atlantis Studies in Dynamical Systems 4, DOI 10.2991/978-94-6239-174-1

[Burc79] BURCKEL, R., *An introduction to classical complex analysis*, vol. 1, Academic Press, New York, 1979.

[Burg11] BURGUET, D., C^2 surface diffeomorphisms have symbolic extensions. *Invent. Math.* 186 (2011), 191–236.

[Buz97] BUZZI, J., Intrinsic ergodicity for smooth interval maps. *Israel J. Math.* 100 (1997), 125–161.

[BS03] BUZZI, J. and SARIG, O., Uniqueness of equilibrium measures for countable Markov shifts and multidimensional piecewise expanding maps. *Ergod. Th. & Dynam. Sys.* 23 (2003), 1383–1400.

[Ca94] CANNON, J.W., The combinatorial Riemann mapping theorem. *Acta Math.* 173 (1994), 155–234.

[CFP07] CANNON, J.W., FLOYD, W.J., and PARRY, R., Constructing subdivision rules from rational maps. *Conform. Geom. Dyn.* 11 (2007), 128–136.

[CG93] CARLESON, L. and GAMELIN, T.H., *Complex dynamics*, Springer, New York, 1993.

[CZ15] CHUNG, N.-P. and ZHANG, G., Weak expansiveness for actions of sofic groups. *J. Funct. Anal.* 268 (2015), 3534–3565.

[Com09] COMMAN, H., Strengthened large deviations for rational maps and full shifts, with unified proof. *Nonlinearity* 22(6) (2009), 1413–1429.

[CRL11] COMMAN, H. and RIVERA-LETELIER, J., Large deviation principles for non-uniformly hyperbolic rational maps. *Ergod. Th. & Dynam. Sys.* 31 (2011), 321–349.

[Con85] CONWAY, J.B., *A course in functional analysis*, Springer, New York, 1985.

[CT11] CUI, G. and TAN, L., A characterization of hyperbolic rational maps. *Invent. Math.* 183 (2011), 451–516.

[CT15] CUI, G. and TAN, L., Hyperbolic-parabolic deformations of rational maps. Preprint, (arXiv:1501.01385), 2015.

[DS97] DAVID, G. and SEMMES, S., *Fractured fractals and broken dreams*, Oxford Lecture Series in Mathematics and its Applications 7, The Clarendon Press, Oxford University Press, New York, 1997.

[DPU96] DENKER, M., PRZYTYCKI, F., and URBAŃSKI, M., On the transfer operator for rational functions on the Riemann sphere. *Ergod. Th. & Dynam. Sys.* 16 (1996), 255–266.

[DU91] DENKER, M. and URBAŃSKI, M., Ergodic theory of equilibrium states for rational maps. *Nonlinearity* 4(1) (1991), 103–134.

[DFPV12] DÍAZ, L.J., FISHER, T., PACIFICO, M.J., and VIEITEZ, J.L., Entropy-expansiveness for partially hyperbolic diffeomorphisms. *Discrete Contin. Dyn. Syst.* 32 (2012), 4195–4207.

[Dob68] DOBRUŠIN, R.L., Description of a random field by means of conditional probabilities and conditions for its regularity. *Theor. Verojatnost. i Primenen* 13 (1968), 201–229.

[DH93] DOUADY, A. and HUBBARD, J.H., A proof of Thurston's topological characterization of rational functions. *Acta Math.* 171 (1993), 263–297.

[Dow11] DOWNAROWICZ, T., *Entropy in Dynamical Systems*, Cambridge Univ. Press, Cambridge, 2011.

[DM09] DOWNAROWICZ, T. and MAASS, A., Smooth interval maps have symbolic extensions *Invent. Math.* 176 (2009), 617–636.

[DN05] DOWNAROWICZ, T. and NEWHOUSE, S., Symbolic extensions and smooth dynamical systems. *Invent. Math.* 160 (2005), 453–499.

[Du10] DURRETT, R., *Probability: theory and examples*, Cambridge Univ. Press, Cambridge, 2010.

[Fo99] FOLLAND, G.B., *Real analysis: modern techniques and their applications*, 2nd ed., Wiley, New York, 1999.

[FK83] FURSTENBERG, H. and KIFER, Y., Random matrix products and measures on projective spaces. *Israel J. Math.* 46 (1983), 12–32.

[FLM83] FREIRE, A., LOPES, A., and MAÑÉ, R., An invariant measure for rational maps. *Bol. Soc. Brasil. Mat.* 14 (1983), 45–62.

[GOY88] GREBOGI, C., OTT, E., and YORKE, J., Unstable periodic orbits and the dimension of multifractal chaotic attractors. *Phys. Rev. A* 37 (1988), 1711–1724.

[GP10] GUILLEMIN, V. and POLLACK, A., *Differential topology*, Amer. Math. Soc., Providence, RI, 2010.

[HP09] HAÏSSINSKY, P. and PILGRIM, K., Coarse expanding conformal dynamics. *Astérisque* 325 (2009).

[HP14] HAÏSSINSKY, P. and PILGRIM, K., Minimal Ahlfors regular conformal dimension of coarse conformal dynamics on the sphere. *Duke Math. J.* 163 (2014), 2517–2559.

[Ha02] HATCHER, A., *Algebraic topology*, Cambridge Univ. Press, Cambridge, 2002.

[HT03] HAWKINS, J. and TAYLOR, M., Maximal entropy measure for rational maps and a random iteration algorithm for Julia sets. *Intl. J. of Bifurcation and Chaos* 13 (6) (2003), 1442–1447.

[He01] HEINONEN, J. *Lectures on analysis on metric spaces*. Springer, New York, 2001.

[HK98] HEINONEN, J. and KOSKELA, P., Quasiconformal maps in metric spaces with controlled geometry. *Acta Math.* 181 (1998), 1–61.

[HSS09] HUBBARD, J.H., SCHLEICHER, D., and SHISHIKURA, M., Exponential Thurston maps and limits of quadratic differentials. *J. Amer. Math. Soc.* 22 (2009), 77–117.

[IRRL12] INOQUIO-RENTERIA, I. and RIVERA-LETELIER, J., A characterization of hyperbolic potentials of rational maps. *Bull. Braz. Math. Soc. (N.S.)* 43 (2012), 99–127.

[KPT15] KAHN, J., PILGRIM, K., and THURSTON, D., Conformal surface embeddings and extremal length. Preprint, (arXiv:1507.05294), 2015.

[KK00] KAPOVICH, M. and KLEINER, B., Hyperbolic groups with low-dimensional boundary. *Ann. Sci. École Norm. Sup.* 33 (2000), 647–669.

[KH95] KATOK, A. and HASSELBLATT, B., *Introduction to the modern theory of dynamical systems*, Cambridge Univ. Press, Cambridge, 1995.

[Ki90] KIFER, Y., Large deviations in dynamical systems and stochastic processes. *Trans. Amer. Math. Soc.* 321 (1990), 505–524.

[KR95] KORÁNYI, A. and REIMANN, H.M., Foundations for the theory of quasiconformal mappings on the Heisenberg group. *Adv. Math.* 11 (1995), 1–87.

[Kü97] KÜHNAU, R., Herbert Grötzsch zum Gedächtnis. *Jahresber. Dtsch. Math. Ver.* 99 (1997), 122–145.

[Li13] LI, Z., Periodic points and the measure of maximal entropy of an expanding Thurston map. To appear in *Trans. Amer. Math. Soc.*

[Li14] LI, Z., Equilibrium states for expanding Thurston maps. Preprint, (arXiv:1410.4920), 2014.

[Li15] LI, Z., Weak expansion properties and large deviation principles for expanding Thurston maps. *Adv. Math.* 285 (2015), 515–567.

[LVY13] LIAO, G., VIANA, M., and YANG, J., The entropy conjecture for diffeomorphisms away from tangencies. *J. Eur. Math. Soc. (JEMS)* 15 (2013), 2043–2060.

[Lo77] LOÈVE, M., *Probability theory*, vol. 1, 4th ed., Springer, New York, 1977.

[Ly83] LYUBICH, M., Entropy properties of rational endomorphisms of the Riemann sphere. *Ergod. Th. & Dynam. Sys.* 3 (1983), 351–385.

[Mar13] MARKOVIC, V., Criterion for Cannon's Conjecture. *Geom. Funct. Anal.* 23 (2013), 1035–1061.

[Man87] MAÑÉ, R., *Ergodic theory and differentiable dynamics*, Springer, Berlin, 1987.

[MauU03] MAULDIN, D. and URBAŃSKI, M., *Graph directed Markov systems: geometry and dynamics of limit sets*, Cambridge Univ. Press, Cambridge, 2003.

[MayU10] MAYER, V. and URBAŃSKI, M., Thermodynamical formalism and multifractal analysis for meromorphic functions of finite order. *Mem. Amer. Math. Soc.* 203 (2010), no. 954, vi+107 pp.

[Mc95] MCMULLEN, C.T., The classification of conformal dynamical systems. In YAU, S.T. et al. (Eds.), *Current developments in mathematics* (pp. 323–360), Internat. Press, Cambridge, MA, 1995.

[Mc98] MCMULLEN, C.T., Kleinian groups and John domains. *Adv. Math.* 135 (1998), 351–395.

[Mc00] MCMULLEN, C.T., Hausdorff dimension and conformal dynamics. II. Geometrically finite rational maps. *Comment. Math. Helv.* 75 (2000), 535–593.

[Mc08] McMULLEN, C.T., Thermodynamics, dimension and the WeilPetersson metric. *Invent. Math.* 173 (2008), 365–425.

[MS98] McMULLEN, C.T. and SULLIVAN, D.P., Quasiconformal homeomorphisms and dynamics. III. The Teichmüller space of a holomorphic dynamical system. *Adv. Math.* 135 (1998), 351–395.

[Me02] MEYER, D., Quasisymmetric embedding of self similar surfaces and origami with rational maps. *Ann. Acad. Sci. Fenn. Math.* 27 (2002), 461–484.

[Mil06] MILNOR, J., *Dynamics in one complex variable*, 3rd ed., Princeton Univ. Press, Princeton, 2006.

[Mis73] MISIUREWICZ, M., Diffeomorphisms without any measure with maximal entropy. *Bull. Acad. Pol. Sci.* 21 (1973), 903–910.

[Mis76] MISIUREWICZ, M., Topological conditional entropy. *Studia Math.* 55 (1976), 175–200.

[Moi77] MOISE, E., *Geometric topology in dimensions* 2 *and* 3, Springer, New York, 1977.

[Mos73] MOSTOW, G.D., *Strong rigidity of locally symmetric spaces*, Annuals of Mathematics Studies 78, Princeton Univ. Press, Princeton, 1973.

[Mu00] MUNKRES, J.R., *Topology*, 2nd ed., Prentice Hall, Upper Saddle River, NJ, 2000.

[Ol03] OLIVEIRA, K., Equilibrium states for non-uniformly expanding maps. *Ergod. Th. Dynam. Sys.* 23 (2003), 1891–1905.

[PV08] PACIFICO, M.J. and VIEITEZ, J.L., Entropy-expansiveness and domination for surface diffeomorphisms. *Rev. Mat. Complut.* 21 (2008), 293–317.

[Pan89] PANSU, P., Métriques de Carnot-Carathéodory et quasiisométries des espaces symétriques de rang un. *Ann. of Math. (2)* 129 (1989), 1–60.

[Par64] PARRY, W., On intrinsic Markov chains. *Trans. Amer. Math. Soc.* 112 (1964), 55–66.

[Pe97] PESIN, YA., *Dimension theory in dynamical systems: contemporary views and applications*, Chicago Univ. Press, Chicago, 1997.

[Pi03] PILGRIM, K., *Combinations of complex dynamical systems*, volume 470 of *Lect. Notes in Math.*, Springer, Berlin, 2003.

[PSh96] POLLICOTT, M. and SHARP, R., Large deviations and the distribution of pre-images of rational maps. *Comm. Math. Phys.* 181 (1996), 733–739.

[PSr07] POLLICOTT, M. and SRIDHARAN, S., Large deviation results for periodic points of a rational map. *J. Dyn. Syst. Geom. Theor.* 5 (2007), 69–77.

[Pr90] PRZYTYCKI, F., On the Perron-Frobenius-Ruelle operator for rational maps on the Riemann sphere and for Hölder continuous functions. *Bol. Soc. Brasil. Mat.* 20(2) (1990), 95–125.

[PRL11] PRZYTYCKI, F. and Rivera-Letelier, J., Nice Inducing Schemes and the Thermodynamics of Rational Maps. *Comm. Math. Phys.* 301 (2011), 661–707.

[PU10] PRZYTYCKI, F. and URBAŃSKI, M., *Conformal fractals: ergodic theory methods*, Cambridge Univ. Press, Cambridge, 2010.

[Ro49] ROKHLIN, V., On the fundamental ideas of measure theory. *Mat. Sb. (N.S.)*, 25(67) (1949), 107–150; English transl., *Amer. Math. Soc. Transl.* (1) 10 (1962), 1–54.

[Ro61] ROKHLIN, V., Exact endomorphisms of a Lebesgue space. *Izv. Akad. Nauk SSSR Ser. Mat.* 25 (1961), 499–530; English transl., *Amer. Math. Soc. Transl.* (2) 39 (1964), 1–36.

[Ru89] RUELLE, D., The thermodynamical formalism for expanding maps. *Comm. Math. Phys.* 125 (1989), 239–262.

[Si72] SINAI, YA., Gibbs measures in ergodic theory. *Russian Math. Surveys* 27 (1972), 21–69.

[SU00] STRATMANN, B.O. and URBAŃSKI, M., The geometry of conformal measures for parabolic rational maps. *Math. Proc. Cambridge Philos. Soc.* 128 (2000), 141–156.

[SU02] STRATMANN, B.O. and URBAŃSKI, M., Jarník and Julia: a Diophantine analysis for parabolic rational maps for geometrically finite Kleinian groups with parabolic elements. *Math. Scand.* 91 (2002), 27–54.

[Su85] SULLIVAN, D., Quasiconformal homeomorphisms and dynamics. I. Solution of the Fatou-Julia problem on wandering domains. *Ann. of Math. (2)* 122 (1985), 401–418.

[Th16] THURSTON, D., From rubber bands to rational maps: Research report. *Res. Math. Sci.* (2016) 3:15.

[TV80] TUKIA, P. and VÄISÄLÄ, J., Quasisymmetric embeddings of metric spaces. *Ann. Acad. Sci. Fenn. Ser. A I Math.* 5 (1980), 97–114.

[Wal76] WALTERS, P., A variational principle for the pressure of continuous transformations. *Amer. J. Math.* 17 (1976), 937–971.

[Wal82] WALTERS, P., *An introduction to ergodic theory*, Springer, New York, 1982.

[Wan14] WANG, X., A decomposition theorem for Herman maps. *Adv. Math.* 267 (2014), 307–359.

[Wi07] WILDRICK, K.M., Quasisymmetric parameterizations of two-dimensional metric spaces. PhD thesis, University of Michigan, 2007.

[XF07] XIA, H. and FU, X., Remarks on large deviation for rational maps on the Riemann sphere. *Stoch. Dyn.* 7(3) (2007), 357–363.

[Yu03] YURI, M., Thermodynamic formalism for countable to one Markov systems. *Trans. Amer. Math. Soc.* 335 (2003), 2949–2971.

[Zh08] ZHANG, G., Dynamics of Siegel rational maps with prescribed combinatorics. Preprint, (arXiv:0811.3043), 2008.

[ZJ09] ZHANG, G. and JIANG, Y., Combinatorial characterization of sub-hyperbolic rational maps. *Adv. Math.* 221 (2009), 1990–2018.

[Zi96] ZINSMEISTER, M., *Formalisme thermodynamique et systèmes dynamiques holomorphes*, Panoramas et Synthèses 4, Société Mathématique de France, Paris, 1996.

Index

Z. Li, *Ergodic Theory of Expanding Thurston Maps*, Atlantis Studies in Dynamical Systems 4, DOI 10.2991/978-94-6239-174-1

Printed in the United States
By Bookmasters